EMPIRE *of the* BEETLE

EMPIRE

of the

ANDREW NIKIFORUK

BEETLE

How Human Folly and a Tiny Bug Are
Killing North America's Great Forests

David Suzuki Foundation

GREYSTONE BOOKS
D&M PUBLISHERS INC.
Vancouver/Toronto/Berkeley

Greystone Books
An imprint of D&M Publishers Inc.
2323 Quebec Street, Suite 201
Vancouver BC Canada V5T 4S7
www.greystonebooks.com

David Suzuki Foundation
219–2211 West 4th Avenue
Vancouver, BC Canada V6K 4S2

Cataloguing data available from Library and Archives Canada
ISBN 978-1-55365-510-7 (pbk.)
ISBN 978-1-55365-894-8 (ebook)

Editing by Barbara Pulling
Cover and text design by Naomi MacDougall
Front cover illustrations: mountain pine beetle © Jesse Koreck;
lodgepole pine © Debra Ward
Tree illustrations by Debra Ward www.debraward.ca
Tree trunk cross-section on page 46 based on an original drawing by Diana Six
Printed and bound in Canada by Friesens
Text printed on acid-free, 100% post-consumer paper
Distributed in the U.S. by Publishers Group West

We gratefully acknowledge the financial support of the Canada
Council for the Arts, the British Columbia Arts Council, the Province of British
Columbia through the Book Publishing Tax Credit, and the Government
of Canada through the Canada Book Fund for our publishing activities.

To landowners who care about
the geography of hope

Contents

. . . . .

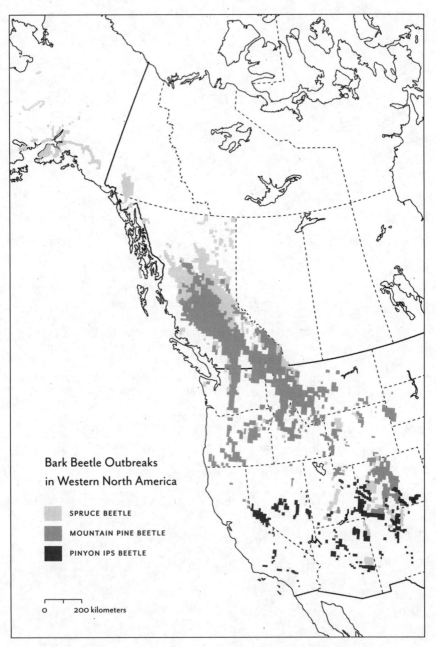

Bark Beetle Outbreaks
in Western North America

SPRUCE BEETLE

MOUNTAIN PINE BEETLE

PINYON IPS BEETLE

0   200 kilometers

Sources: Canadian Forest Service, British Columbia Ministry of Forests, and U.S. Forest Service

"You can plan all you want to. You can lie in your morning bed and fill whole notebooks with schemes and intentions. But within a single after-noon, within hours or minutes, everything you plan and everything you have fought to make yourself can be undone as a slug is undone when salt is poured on him. And right up to the moment when you find yourself dissolving into foam you can still believe you are doing fine."

WALLACE STEGNER, *Crossing to Safety*

"Now, it never seems to occur to these far-seeing teachers that Nature's object in making animals and plants might possibly be first of all the happiness of each one of them, not the creation of all for the happiness of one."

JOHN MUIR, *A Thousand-Mile Walk to the Gulf*

"We must learn to grow like a tree, not like a fire."

WENDELL BERRY, *Sex, Economy, Freedom & Community*

IN THE late 1870s, swarms of locusts erupted throughout the North American West, turning the skies black and leaving fertile valleys "as bare as winter." The Rocky Mountain locust (*Melanoplus spretus*) poured such a carpet of insects onto the prai-rie that oil from their crushed bodies stopped trains dead in their tracks in Texas. One swarm, perhaps the greatest concentration of animals on earth, consisted of approximately 3.5 trillion insects weighing 1.7 million tons. The swarm extended over an area 1,800 miles long and 110 miles wide.

Shocked and awed, pioneers described the "rasping whirring" of locusts as their wings "glistened and glittered" in the sun. The working of the insect's jaws made such a clattering that it reminded some of a crackling prairie fire. The bugs fell haillike on rooftops, and "every blade of grass bore its burden." After one Kansas farmer watched the creatures descend upon on his sixty-acre wheat crop and vegetable garden like hungry, demonic smoke (everything was eaten within an hour), he wrote that "no man can successfully fight against nature."

By 1878, flights of the Rocky Mountain locust had leveled most grain fields in Colorado, Manitoba, Kansas, Minnesota, and Missouri. In some counties, entire families of people starved. Others learned to eat the locusts, minus the legs and wings. When they ran out of crops, grass, and trees, the famished insects ate leather stirrups, ax handles, clapboard, sheep's wool, and farmers' laundry. And then the hordes crashed and fell silent.

The "locust evil" prompted scientific investigations in Canada and the United States as well as two congressional reports. The U.S. government eventually recommended increased settlement and cultivation, combined with massive irrigation, to check the "disastrous incursions of these devouring pests." By dumb luck and destructive land engineering, pioneers overwhelmed the locusts. Once the grasslands where the insect nested and laid its eggs were gone, the species went extinct.

In the late 1980s, a series of bark beetle plagues exploded across the West with locust-like ferocity. The beetles, mostly *Dendroctonus ponderosae* and *Dendroctonus rufipennis*, attacked conifers in swarms so large that they appeared on local airport radar to be rainstorms. Some beetle flights caught updrafts and crossed the

Rocky Mountains, traveling distances of over 175 miles. What some called the "Katrina of the West" attacked mature forests and young plantation trees until there was nothing left for the bark borers to eat. By 2010, the insect had girdled and killed more than 30 billion lodgepole, pinyon, ponderosa, and whitebark pines as well as white spruce and Engelmann spruce. Human loggers destroyed almost as many trees in a vain attempt to stop the invasion. It may well be the greatest forest die-off recorded since industrious peasant communities deforested much of northern Europe between the eleventh and thirteenth centuries.

After the beetle swarming, dead trees fell on houses, blocked roads, and severed power lines. By collapsing entire forests, the bark beetle affected watersheds, fire regimes, carbon balances, local temperatures, and the livelihood of countless logging communities. Although climate change triggered the completely unprecedented event, human ignorance and arrogance set the stage for the second-largest insect outbreak in North American history. And this time around, human intervention has failed to contain the plague because manmade landscapes largely drove it. With little warning and even less fanfare, the West, the land of sage and mountain light, the place Wallace Stegner once called "the geography of hope," has fallen to the empire of the beetle.

ONE

The Alaska Storm

. . . . .

"You get tragedy where the tree, instead of bending, breaks."

LUDWIG WITTGENSTEIN, *Culture and Value*

THE FIRST of the continent's great beetle outbreaks overran Ed Berg's backyard outside Homer, Alaska, in 1993. That's when spruce beetles hit the towering ancient old-growth Sitka spruce on the ecologist's one-acre lot just off East End Road. "You could hear a little crunch as they were going into the bark," recalls Berg. Within a year or two, about 95 percent of his trees turned red, dropped their needles, and died. So, too, did the trees at his neighbor's place. Most people had built off East End Road to enjoy the splendor of, say, a 270-year-old Sitka spruce. Now the trees stood as lifeless as gray skeletons. "We were all heartbroken," says Berg. "It was a gut-wrenching experience."

Berg's front-row seat to the epidemic didn't end with his chain saw. As ecologist for the Kenai National Wildlife Refuge, a 2-million-acre playground for moose, salmon, and spruce trees on Alaska's south-central coast, Berg watched the beetle chew its way through an area twice the size of Yellowstone National

Park, killing virtually all of the mature spruce trees, be they white, Sitka, or the white-Sitka hybrid called Lutz spruce. Much to Berg's dismay, the refuge became home to the largest spruce beetle outbreak ever recorded in the world. "We had Alfred Hitchcock–style in-your-face beetle flights. It was phenomenal," he says. To this day, government officials refer to the death of more than 200 million trees over 4 million acres as "the largest natural disturbance" to unsettle south-central Alaska since the 1964 earthquake. Some have even called it Alaska's tsunami.

GIVEN THAT white spruce is the most common and productive tree in the boreal forest (it makes fine paper as well as great canoe paddles and piano sounding boards), any massive die-off should be cause for concern. The boreal, a sort of poor person's Amazon, covers nearly 15 percent of the earth and does a lot of oxygen making and carbon storing. Although the deaths of tens of millions of trees invited both grief and reflection in Alaska and the Yukon, most media ignored the event. The *New York Times* didn't appreciate the significance of the storm till long after it had passed. Yet Alaska's strange adventure with the spruce beetle foreshadowed a series of unfortunate events that would change landscapes all across the West. Alaska was but the first of several beetle conquests that collapsed aging pine and spruce forests from British Columbia to New Mexico. Ten years after the Kenai storm, Utah forestry officials sounded just like Alaskans when they described how spruce beetles had taken a 300,000-acre bite out of the Dixie National Forest outside Salt Lake City: "It's really something to see. You would be very surprised. It's hard to describe until you see it. It's just dead trees as far as the eye can see."

The Kenai Peninsula, known as "Alaska's Playground," embodies the independent and conflicted spirit of the modern West. Here an artist can produce a painting titled *dissin' the house that narcissus built*. An eccentric elderly lady can feed hundreds of bald eagles with scraps from the local fish plant. Most people carry guns as both a right and a moral responsibility. After John Muir visited the peninsula and other pleasant Alaska spots in the nineteenth century, the conservationist thought that he had reached "the very paradise of the poets, the abode of the blessed."

That's certainly how Berg and most other residents feel. The "bad disease" known as Alaska grabbed hold of the Wisconsin philosophy major in 1977 and never let go. (Nor did the philosophers: Berg's favorites remain Ludwig Wittgenstein and some of the "old Greek guys.") Berg, who dresses in suspenders and flannel shirts, loved the elegance of the old trees and the creatures that thrived under their spacious canopy. After he'd worked as a carpenter and a geology teacher in Alaska for several years, his love of plants took him back to school. He completed his PhD in botany at the University of Georgia in 1993 and returned home to work on the refuge, where the U.S. Fish and Wildlife Service wanted someone who could look at the "big picture." That year the big picture included 300,000 acres of fresh beetle kill. Berg never bargained that a beetle the size of a match head would frame the picture and eventually change everything in it. Ever since then, he's continued to explain what happened and why with the patience of a forensic scientist. Even the U.S. Congress has quoted his work.

Until the outbreak in Alaska, most experts thought the spruce beetle, *Dendroctonus rufipennis*, was a relatively humble and mild-mannered creature. The beetle manages spruce forests everywhere from Newfoundland to Alaska. It typically targets weak trees or older, slow-growing ones that can't make much resin. As a result, it directly competes with the forestry industry for mature large-diameter timber. That's why Stephen L. Wood, the Mormon who wrote the bible on bark beetles, described *rufipennis* as "the most destructive of the spruce inhabiting bark beetles." Unlike many of its relatives, *rufipennis* can also hibernate

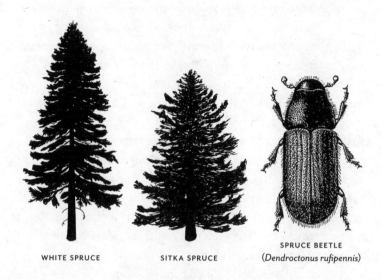

WHITE SPRUCE　　　　SITKA SPRUCE　　　　SPRUCE BEETLE
(*Dendroctonus rufipennis*)

in a unique way. Newly formed adults that emerge in the fall simply overwinter at the base of the tree under the bark. According to Wood, the practice "places overwintering beetles below the snow line, where they are protected from extremely low winter temperatures and predators such as woodpeckers." It's a spruce beetle specialty. Another is its fondness for wind-blown timber. Recently downed white spruce makes an undefended chemical-free nursery. In fact, more than ten times as many spruce beetles can thrive in windfall as in a green, well-defended tree.

Between 1939 and 1953, a massive spruce beetle outbreak grayed forests in the central Rocky Mountains. A heavy windfall started things off. "While protected from their natural enemies, the woodpeckers, by snow that covered the windfall spruce, large populations of insects developed and spread to surrounding green trees," reported Fort Collins foresters in 1956. "This is typical of the way in which infestations develop." It remains so.

During the Alaska outbreak, Berg taught locals how to spot the difference between *Ips perturbatus*, a minor tree mugger, and *Dendroctonus*, the tree slayer, by pulling out a magnifying glass and a penknife. The rear ends on these "creatures of disturbance" look as different from one another as those on antique cars. The *Ips*, or engraver beetle, has a spiny rear end that looks like "the finned taillights" on a 1950s Pontiac or pink Cadillac. In contrast, Berg says, *Dendroctonus* "has a nice, well-rounded rear end, reminiscent of, say, a '49 Ford." Identifying a Pontiac on their spruce trees gave landowners a sense of peace, whereas locating a Ford meant hauling out the chain saw. The piles of sawdust left by the *Ips* beetle, which dines on the warm, sunlit tops of downed tree trunks, was no cause for alarm, wrote Berg. But once residents found a parked *Dendroctonus*, "you could pretty well kiss the rest of your forest goodbye."

The U.S. Forest Service had generally rated the spruce beetle as a marginal player in southern and central Alaska's spruce woodlands. Whenever it erupted in mature stands, the beetle killed no more than 30 percent of the spruce trees. Moreover, a beetle boom hardly ever lasted more than three years before cold winters and tree resistance snuffed it out. The creature also took two long years to reach maturation. Most entomologists, such as Ed Holsten with the U.S. Forest Service, considered the bug "a secondary actor." Although the state began "red needle" aerial surveys in the 1970s, no one thought *Dendroctonus* was capable of a "holy shit" event, especially in a cool, wet place like the Kenai.

The epidemic probably began in a pile of blown-down trees in the mid-1970s, a few years before Berg moved to Alaska. "It's quite a salad bowl for beetles up here. We have very few hardwoods," explains Holsten. The entomologist spotted a swath of red, attacked trees about 50,000 acres in size during a routine forest pest aerial survey just south of Anchorage in the early 1990s. "The beetles got up to a critical mass, and then they shot up into a

real outbreak and went on and on," says Holsten. "It caught everyone by surprise."

By 1986, the outbreak was chewing its way through 100,000
acres a year and was moving south and along the coast. Windstorms produced more beetle nurseries throughout the region,
the same way lightning strikes create a variety of fires. In the
beginning, most Alaskans couldn't believe that a small beetle
could take down a 200-year-old, 125-foot-high Sitka spruce more
than 20 inches in diameter. "You kind of expect something the
size of a guinea pig or German shepherd out here killing all these
trees," Holsten told the *Anchorage Daily News* in 1986. He held up
a beetle the size of a rice kernel and said, "People look at this and
say, 'That's it?'" Locals routinely brought Holsten two-inch-long
cerambycid beetles during the epidemic, thinking that these
large wood borers were doing the killing. "Well, no," he'd explain.
"It's a little thing less than an eighth of an inch long."

The Anchorage newspaper made an astute observation about
the infestation: "In a sense the beetle is managing spruce forests by killing the old trees and leaving the new. But the Forest
Service would rather manage its own trees. And has been experimenting with ways to beat the beetle at its game." That included
thinning trees, spraying pesticides such as lindane and carbaryl,
and hanging up pheromone traps. Holsten predicted that a cold
winter might stop the beetle outbreak dead. But that didn't happen. More hillsides throughout the peninsula started to turn red
and pink as the beetles bored into more mature spruce.

During the late 1980s, the beetles continued to emerge in
unbelievable numbers. Skeeter Werner, a research entomologist
and a colleague of Holsten's, watched clouds of bugs fly in May
and June from one valley to another in the northern Kenai. "On
warm days in June, a cloud of beetles would go up a valley in the
late afternoon, and the wind would carry them over a mountain.
It was amazing." (Sometimes wind currents carried the bugs

as far north as Barrow, Alaska, where the Inuit had no name for them.) An incredulous Werner noted that the beetles often killed 80 to 90 percent of the trees. He even found beetles attacking spruce trees as small as three inches in diameter. "We didn't know anything about them, really. We couldn't predict how they would behave, and they dispersed so rapidly. It was staggering."

The Forest Service thinned some stands of trees at first to slow down the beetle advance. According to Werner, "It worked fine for a year, but the beetle population was so great that they wiped out the thinned stands." Some foresters then called for controlled burns to stop the beetles. "But we couldn't get the approval of the EPA," says Werner, who routinely kept containers of beetles in his home refrigerator for research purposes. (His wife often found beetles crawling around on the floor.) Because so many different groups, agencies, and municipalities owned land on the Kenai, everything from federal parks to Native corporations, no real consensus emerged on what to do. Finally, the experts went to the media. "By that time it was too damn late," says Werner.

Holsten, who retired in 2005, now remembers 1993 as "the mother of all summers." For the first time, a majority of the beetles completed their reproduction cycle in just one year instead of the normal two. That spring, both one-year-old and two-year-old adults had bolted from the trees, a development that doubled the number of beetles on the landscape. "They were even scooping beetles off the beach in Mason jars," recalls Holsten. Clouds of beetles flew into people's homes in such great numbers that many residents temporarily fled. Tornadoes of beetles swirled around Holsten's head at forest monitoring plots, getting into his hair and shirt collar. "It was like a swarm of locusts in East Africa."

The massive clouds of beetles that moved over the southern Kenai during the 1990s still haunt people's imaginations. One Homer resident later recalled the absolute otherworldliness of it all: "I can remember the first few years driving in my car and having to use my windshield wipers because of the hordes of beetles. It's almost like how someone would envision a nuclear event where you have dust falling in the air . . . You have this sense that something horrible is coming and on the horizon. Then you watch things turn all brown and turn gray and die and eventually fall over and blow down." Ed Berg remembers meeting a man who described an unusual sighting in the Anchor River valley. "One day he saw a cloud several miles up the valley. He'd never seen anything like that before. The cloud came down the valley and engulfed him. It was a massive beetle flight."

By 1996, almost everyone in the Kenai had a story about beetles, dead trees, or falling beetle kill in their backyard. In an article entitled "Elegy to a Spruce Forest: A Cemetery of Sorts," Homer resident Marie McCarty vividly described how clouds of beetles invaded her property one warm May day. The swarm "tangled in my long hair, whirred past my sunglasses, spun suicidally in a Behr Oil can. They landed on my children's sandwiches. The kids squished as many of them as they could after we coached them with a little forest ecology lesson." She watched "the trees bleed amber pitch through ladybug-sized holes" and heard the beetles killing spruce trees in stereo. They "made a sound like granular snow melting in a forest on a sunny day, but sinister."

By the time Berg had bucked up and removed nearly 130 dead trees from his own land, the ecologist had realized that something phenomenal was going on in the wildlife refuge and beyond. The beetles took no prisoners and executed between 80 and 90 percent of the mature trees "Mafia-style." That was unusual, and the scale also seemed out of whack. Between 1993 and 1994 alone, the area of forest under active attack grew from 600,000 to 800,000 acres. Berg suspected that he was witnessing a natural event greatly exaggerated and enhanced by climate change, though he didn't have any evidence at the time to prove either point.

He started reading about epic bug battles in foreign lands. "I got hold of some Norway publications, because they've been dealing with the spruce beetle for hundreds of years and tried various things," he recalls. Although the Norwegian studies concerned *Ips typographus*, a beetle that swarms damaged, drought-stressed Norwegian spruce, the species has many similarities to *Dendroctonus rufipennis*. The two beetles look remarkably alike; they transport similar numbers of hitchhiking mites and carry a variety of fungi for nutrition, including some of the most potent blue-staining fungi on the planet. Whereas *Ips* males direct a tree attack, *Dendroctonus* females lead the swarm.

The Norwegian experience read like a dark yet familiar fairy tale, Berg found. Powerful windstorms combined with extreme drought had created major epidemics in the country in the 1840s and again in the 1970s, with some able-bodied assistance from the logging industry. Throughout the 1960s, spruce loggers stored cut trees in huge open piles next to forests without debarking them. To a hairy eight-toothed spruce beetle, a fresh-cut spruce log looks just as appetizing as a wind-downed picnic. As a consequence, the logging industry cultivated high beetle populations in the woods. After a storm blew down thousands of trees in southeastern Norway in 1969, the beetles took off, killing more than 4 million trees over an area roughly the size of Florida. At the time, Norway was recording its driest and hottest summers since 1840. The countryside received only a third of the normal rainfall, resulting in low water tables and drought-stressed spruce. The nasty dry spell fueled the epidemic well into 1982.

The Norwegians tried just about everything imaginable to stop the tree killing. In one county alone, they logged 300,000 infested trees, but the beetles kept on coming. The government then nailed up pipe traps that looked like odd-shaped pagodas and baited them with chemical pheromones to attract male beetles. But the traps killed *Dendroctonus's* most efficient predator, the clerid beetle. Chastened scientists modified the entrance to the contraptions in order to spare the clerid. In 1979, the government tacked more than 600,000 traps onto spruce trees throughout southern Norway. Each contraption caught an average of 16,000 beetles, for a total annual catch of 10 billion.

Saturday-night TV hosts in Norway chatted wittily about the funny traps lining hiking trails. Newspaper cartoonists imagined bark beetles as fierce resistance fighters armed with machine guns. The rock song "Bark Beetle Boogie" hit the top of the Norwegian charts. (You have to be Norwegian to appreciate the lyrics: "We are a thousand bark beetles in every bark beetle tree and dance bark beetle boogie until the tree falls down.") But the mass trapping, which cost tens of millions of dollars, couldn't stop the beetle boogie. The outbreak ended only when the beetle ran out of food.

Unlike Norway, the Kenai didn't have records going back three hundred years. In addition, the wildlife refuge had never been logged. "So we had no idea what the history of the beetle was here, or whether it was new or part of the natural cycle of things," says Berg. As people asked more questions about fire risks and what the refuge was going to do about the outbreak, Berg decided to solve the riddle. He didn't think clear-cutting the region made much sense if the beetle was only doing what it was meant to do: renewing an aging forest.

He returned to the library to learn about tree rings before heading for the woods. Two studies on beetle outbreaks in Engelmann spruce forests in the Colorado Rockies caught his attention. They described how trees the size of fence poles had survived the attacks. After the older trees died, the young ones grew faster, producing much larger tree rings than normal. "Released" from competition, the young survivors surged in size the way well-fed teenagers do. Berg proposed a similar tree-ring study in the Kenai to determine if the beetle was a natural part of the landscape or some sort of "invasion from outside," like chestnut blight or Dutch elm disease. If growth rings in local spruce stands showed ecstatic bursts, then the spruce beetle in the Kenai was as Alaskan as a moose.

Between 1994 and 2002, Berg and a team of helpers cored hundreds of trees from 23 stands of white, Sitka, and Lutz spruce scattered around the peninsula and across Cook Inlet. After Berg sanded the cores to make the rings visible, he measured their width using a microscope equipped with a sliding micrometer attached to a computer that matched up fat and thin rings. (A tree with 200 rings takes about 20 minutes to measure.) Sure enough, Berg discovered tree rings that showed big bursts of growth after repeated beetle thinnings over 250 years. The biggest growth spurts followed beetle outbreaks roughly every 50 years, in the 1810s, 1880s, 1910s, and 1970s. Berg even cored an old beetle-scarred spruce near Homer that was leaning against a young tree. A swarm of *Dendroctonus* had invaded the tree in 1884, he discovered, leaving behind a wealth of maternal galleries. "We often say that this 1884 beetle-killed tree is as close to the smoking gun of direct evidence as we have gotten in our investigations of bark beetle history," Berg explains. *Dendroctonus* was no tourist.

Berg next cross-checked his tree-ring data against the historical record. About twenty years after the 1880s outbreak, the U.S. government had sent William A. Langille, a northern gold digger and adventurer, to the region to assess its suitability as a forest reserve. In 1904, Langille, who subsequently became known as the father of Alaska forestry, found a hellish and impoverished forest of "old logs and decayed stumps of large size" near Homer. He assumed a fire had created the mess sometime before the Russians occupied Alaska. (To the untrained eye, a beetle-killed forest can look like a burned landscape.) The bearded conservationist John Muir came to exactly the same conclusion during his Alaska travels. Muir speculated that the dead forests near Kachemak Bay had been silenced by volcanic ash. But Berg suspects that both Langille and Muir had poked their heads into the wake of a beetle boom.

The tree-ring data told Berg that the beetle outbreaks had "happened before and will happen again." It also proved that the insect was a true Kenai native that had carefully renewed patches of the spruce forest over a long period. Berg's findings probably played a critical role in dampening support for proposals to clear-cut green timber on refuge and wilderness lands to stop the beetle's progress. "No, we are not going to log off a wildlife refuge due to a beetle outbreak," said Berg at the time. "We're a wildlife refuge, not a timber source."

Canadian loggers had actually tried to outrun a spruce beetle epidemic in the early 1980s. It didn't end well. But the crazy effort did create a patch of logged real estate now visible from outer space. The outbreak, driven by beetle-friendly weather, eerily foreshadowed Kenai's beetle outbreak. In 1975, a serious windstorm followed by a couple of warm summers opened the door to a spruce beetle explosion in the Bowron Lakes area southeast of Prince George, British Columbia. When forestry officials discovered the outbreak three years later, they panicked. In an attempt to control the beetle as well as attendant wildfire risks, they granted loggers permission to clear-cut approximately 125,000 acres in the middle of an important salmon watershed. The idea was to log green trees that had been attacked and thereby deprive the beetle of fuel. At the height of the boom, more than 700 logging trucks delivered green and dead spruce trees to mills in Prince George every day. Over a decade, logging trucks pulled out 175 million cubic feet of wood, enough logs to create a convoy more than a thousand miles long. All in all, the clear-cut yielded enough lumber to build 900,000 homes about 1,200 square feet in size. It took over eight years to restock the clear-cut. More than 6 million seedlings were planted, making the area the "largest contiguous spruce plantation in North America."

Foresters later argued that they had "chased beetles" until

none were left. But in reality, the logging extended the length of the outbreak as well as the size of the clear-cut. Because industry felled the trees in winter, above the snow line, it left massive stumps chock-full of overwintering beetles. The more stumps and slash that loggers left behind, the greater the number of beetle broods that flew through the remaining forest. An outbreak that might have lasted a couple years instead burned away for nearly a decade. And the forest has yet to recover from the logging frenzy. Twenty-five years after the ill-advised attempt to out-log the beetle, scientists still rate most streams in the Bowron as "high risk" or "not properly functioning." In the name of *Dendroctonus*, industry and government sacrificed both fish and water.

A miniature logging boom on the Kenai produced similar results near Ninilchik. After the beetles had finished their work, loggers clear-cut the watershed. Then a flood devastated Deep Creek and the Ninilchik River. According to one fisherman, the flood took "all the fish and the whole river and chucked it into the ocean. It damaged all the holes that the fish came up in and spawned. There were beautiful, beautiful pools. I could count 66 salmon in one hole, and now they're gone."

Nevertheless, because a beetle-killed spruce doesn't retain its saw timber value for much longer than three years, industry in the Kenai pressed for more roads, more clear-cuts, and more logging on federal lands. Between 1991 and 1998, the value of wood sales rose from $2 million to $26 million. On some parts of the peninsula, loggers hastily removed 80 percent of the forest cover. A steady stream of logging trucks took much of the beetle kill to Homer, where it was chipped and loaded onto cargo ships bound for Asian paper markets. But the boom didn't last long. As a result of falling pulp prices, the chipping plant closed by 2004, and most of the loggers went home. Grass, shrubs, and new subdivisions filled in many of the clear-cuts. Real estate agents boasted that treeless landscapes offered an alluring "emerging view."

Ed Berg later argued that the suburban invasion of beetle-killed forest land was "a much bigger ecological response than the trees dying. It's permanent change in the landscape."

During the mid-1990s, the beetles broke all records, killing about 30 million trees a year. Midwestern tourists who flew over the forest started to ask what people called "those beautiful rust-colored trees" on the peninsula. The growing presence of ghostly spruce on the landscape also led to raucous debates about fire and public safety. Citizens packed as many as one hundred public meetings on what to do about beetle kill. Much to the frustration of the U.S. Forest Service and other agencies, no one could agree on a plan. Some communities wanted to log. Others voted to let the dead trees rot naturally. Whenever Holsten recommended salvage logging or systematic thinning, he remembers, audiences would invariably reply, "You're blowing smoke. You just want to cut trees." "We didn't have a good answer to that," he says.

Most people didn't recognize the full impact until the beetle had passed through their community, adds Holsten. "They'd see beetle-killed trees across a river and say, 'They'll never cross the river.' And I'd tell them these beetles can fly up to seven miles. And then they'd cross the river."

In 1998, a spruce beetle task force finally put together "An Action Plan for Rehabilitation in Response to Alaska's Spruce Bark Beetle Infestation." That led to workshops on "hazard tree liability" and forums on the "wildland urban interface." That's jargon for what fire expert Stephen Pyne calls "spark and sprawl," the strange propensity of westerners to build subdivisions in habitats that eventually renew themselves through fire. Researchers also started a group called the Interagency Forest Ecology Study Team, or INFEST, to study the beetle's impact. Government spent tens of millions of dollars on cutting down dead trees around schools, roads, campgrounds, power lines, and 245 miles of rights of way. The Kenai Peninsula Borough's Spruce

Bark Beetle Mitigation Program still publishes annual reports on progress related to "mitigating wildfire and other hazards related to the spruce bark beetle."

To this day, the sheer number of beetles that overwhelmed the forest defies rational calculation. The experts say that trillions upon trillions flew like wobbly WWI aircraft through the forest and leave the figure at that. But here's some math. An average beetle weighs about two-thousandths of an ounce. Given that there are 32,000 ounces in a ton, and that as many as 10,000 spruce beetles can infest just one tree, approximately 19,800 tons of beetles swarmed around the Kenai for a decade. That's like encountering a pod of 3,300 six-ton killer whales swimming through a forest, or finding a mega wolf pack composed of 500,000 eighty-pound members.

During the "mindboggling" outbreak, Werner and Holsten tested the mettle of job applicants for nitty-gritty beetle work by serving them a bottle of beer alongside a plate of sautéed spruce beetle larvae. Technicians repulsed by the beetle meal were low priority for the job of counting or monitoring beetle doings. "The larvae taste like pine nuts and are very tasty, but you have to sauté them in butter," advises Holsten. Eating hard-cased adult beetles would be another matter altogether. "You'd spend all your time flossing your teeth."

Berg, for his part, had started to look at temperature's impact on the beetle. Although *Dendroctonus* had flourished before in the region, it had never killed 30 million trees a year. That was freakish and exceptional. The beetle had consumed twice as much forest in the last twenty years as it had in the previous century. To explain why, Berg examined climate records. Since 1977, he found, most of Alaska had started to warm significantly, at twice the U.S. average rate. He originally thought this might be an Alaskan phenomenon, but reports from two thousand weather stations around the world confirmed that 1977 was some kind of turning point. Between 1977 and 2009, the mean annual temperatures in the Kenai increased by 3 degrees Fahrenheit. Average temperatures shot up in December and January by five and seven degrees respectively. The number of summer days above sixty degrees, the minimum temperature for a beetle flight, expanded from fifty-six to eight-six. "Between 1987 and 1998, we had the longest run of warm summers we've had in several hundred years," says Berg. When he paired temperature records with the amount of acreage killed by the beetle, the two sets of figures clearly tracked each other.

The warming, of course, affected not just beetles but the entire wildlife refuge. Kenai's glaciers, which had retreated by two and a half miles over the last century, continued to melt faster than ice cream. Lake levels dropped by several feet, and kettle ponds turned into patches of grass. Peatlands that had sustained wet moss for nearly twenty thousand years suffered invasions by black spruce and woody shrubs. Gnarled old hemlock trees in alpine country abruptly shot up straight and sober like state troopers. Higher temperatures also invited spruce trees to visibly climb up the slopes of the Kenai Mountains. Locals told Berg that the tree line had moved several hundred feet since the 1940s. And on it went.

The concentrated warming gave the beetles some serious advantages. First, it stressed the hell out of mature spruce trees throughout the region by drying the landscape. (Tree-ring growth has slowed dramatically over the last two decades.) Berg calculated that the available water on the Kenai Peninsula dropped by half. With rising temperatures, three inches of water had simply evaporated into Alaska's blue skies. A landscape's losing half its water is akin to a family's income dropping by 50 percent. By 1995, three years after the beginning of the drought, the volume of red-needle acreages had grown exponentially.

Second, the warming allowed the beetles to do their thing to more lethal effect. In late spring, sugar-hungry trees open their needle pores to breathe in CO_2 for photosynthesis and inadvertently release water vapor. Frozen soil prevents the trees' plumbing from replacing that outgoing water, and trees without water can't mobilize resin. The spruce beetles were hitting the trees in late May—just "when the trees have their pants down," says Berg.

Last but not least, as noted, warmer temperatures allowed the majority of beetles to mature into adults in one year instead of two. This so-called accelerated graduation put more than one generation of beetles in flight every year, and "releasing two generations of adults simultaneously doubles the beetle population," Berg wrote. He concluded that the rising number of warm days combined with a beetle population explosion in a drought-stressed forest to create "the killer punch that brought the southern Kenai Peninsula spruce forests to the mat in 1995 and 1996." The beetles stopped only when they had eaten themselves out of house and home.

Berg's thesis was considered radical in some circles in 2002, but not among ordinary Alaskans. His conclusion that climate change had driven the outbreak seemed obvious. "I had one neighbor who told me that they had just needed someone like me with a PhD to come along and make it official," says Berg. Since then, other researchers have confirmed his conclusions or documented the same pattern in other beetle outbursts. A 2005 study of spruce beetle outbreaks in Colorado and Utah showed that warm winters, hot summers, and drought had set the table for a beetle banquet in mature forests throughout the Rocky Mountains. It's almost as though beetles have a duty to finish off trees crippled by aridity. "We are caught in a rising tide," Berg has written. And that tide is climate change.

AS THE beetles petered out on the Kenai Peninsula, falling and rotting trees became a growing nuisance. The trees crunched homes, smashed camping trailers, blocked roads, and downed

power lines. Businesses such as the Small Potatoes Lumber
Company couldn't keep up with requests to cut down dead trees,
and during windstorms residents could hear the whoosh of their
beetle-killed trees tumbling around their homes like felled giants.
One person described the piles of jackstrawed wood in his yard
as messier than a bombed-out neighborhood in Beirut. Hik-
ers clawed their way through anarchic trails spidered by dead
trees. *Backpacker* magazine described a trek into Kenai's balmy
forests as a walk into "the epicenter of global warming" and
"Camp Apocalypse." Every gusty day brought another power
outage on the peninsula. (In Colorado, Wyoming, and South
Dakota an average of 100,000 beetle-killed trees now fall down
every day.)

Many of the beetle-killed trees snapped off several feet above
the ground due to rot in the sapwood. That caught Berg's attention, because it seemed unusual. He had inspected trees killed by
beetles back in the 1970s, and they had mostly remained standing. Tim Smith, a former logger, told Berg that beetle-killed trees
he'd logged after a major 1974 blowdown in Cooper Landing
showed no evidence of sapwood rot. But the new beetle-killed
wood was rotting away much faster, from the outside in. Berg
concluded that bark beetles had probably carried a whole mess of
rot fungal spores into the trees and warmer summers had accelerated their growth. "Let's hear three cheers for the wood rot fungi!"
Berg wrote in one of his *Refuge Notebooks*, since the rot decreased
the risk of fire.

After several fires ripped through other pockets of beetle
kill in 1996 (the Cripple Creek fire burned through 17,000 acres
near Ninilchik), much beetle talk focused on public safety. Local
politicians hastily put together firewise plans, and landowners
cleared fire zones around their homes. Most forestry officials
assumed that a beetle-killed forest would burn faster and hotter
than normal. Seasoned Alaskans, however, doubted those claims.

Some noted that dead spruce burned "like cardboard" while green ones "burn like Blazo" (white gas).

The fire controversy prompted Berg to reexamine the beetle record in his tree-ring data. He found no charcoal in cores of wood taken from trees dating back to the 1700s. Most of the cool, wet region of the southern Kenai hadn't seen a wildfire for four or five hundred years, in fact, even though beetles routinely thinned the forest every fifty years. In a 2001 newspaper column, Berg wrote that drought, not beetle kill, created the real fire hazard: "A drought-stressed live spruce next to your house is every bit as flammable as a beetle-killed spruce."

Nevertheless, calls for the removal of dead trees exploded. Senator Ted Stevens warned, "The amount of beetle kill forests in Alaska equals a 2.33-mile-wide strip from Washington, D.C., to Los Angeles. With over half of our state's population surrounded by beetle kill forests, it is critical that we find a way to reduce the fuel loads." Senator Frank Murkowski echoed the sentiment, describing Alaska's beetle kill as "an area larger than the entire state of Connecticut, leaving 30 billion board feet of rotting wood just waiting for a match to ignite a giant wildfire."

In 2002, the U.S. Forest Service hired Courtney G. Flint, a geographer and sociologist from the University of Illinois, to figure out why rural communities responded so differently to the beetle. Over two years, Flint interviewed one hundred citizens on the Kenai Peninsula in communities with names like Anchor Point, Moose Pass, and Cooper Landing. "The forest service saw it as a fire management issue and couldn't understand why people weren't on board. They had a one-size-fits-all approach," she says. But what amazed her was the overwhelming collective grief people expressed at the loss of their forest. People who had once dwelled in an old-growth forest like elves inhabited a fire-prone grassland full of stumps. "The beetle challenged and uprooted their sense of identity."

The outbreak also changed the quality of everyday life. In Anchor Point, people were depressed by the number of dead trees and said they now lived in a stump farm: "There are all kinds of kickback from one little beetle." In Ninilchik residents told Flint that "socially we see our neighbors now. We live on a semi-prairie compared to what we lived in before." Hot tub enthusiasts had to give up skinny-dipping, because they'd lost their forest curtain. In Moose Pass, many residents thought the government's beetle outbreak response, with its multi-million-dollar fire and salvage logging plans, was more damaging than the beetle itself. Said one, "It's just something that naturally occurs. You can't stop it. It's Mother Nature doing her thing. The solution to me is to let it go."

Flint remembers in particular one elderly couple from Ninilchik who had moved there from New York to be in the woods. They lost more than two hundred trees during the outbreak and paid a local contractor seven thousand dollars to clean up the deadfall. "Birch trees are the only things we have left," they told Flint with tears in their eyes. "What caused the spruce beetle thing in my opinion, and I'm no scientist, but global warming. The temperature has come up. Oh yes, the temperature has come up."

WHILE ALASKANS debated the finer points of lowering fuel loads in "wildland-urban interface," a different spruce beetle outbreak was rattling the Yukon, right next door. *Dendroctonus* had unsettled an even-aged natural monoculture of white spruce forest in Canada's famed Kluane National Park. The World Heritage Site largely consists of icefields, glaciers, and some of Canada's highest mountains. Several of its splendid river valleys boast dense forests of white spruce. Given all the snow and ice, no one ever expected a beetle outbreak in the park. Winter temperatures usually drop to –50 degrees Fahrenheit for at least a week, which guards the forest. Canada's forest service didn't even bother to reconnoiter the area for bark beetles. But after a string of warm winters and dry summers, Kluane became new beetle territory.

The invasion, which trailed Alaska's outbreak, has persisted. Rod Garbutt, a technician with the Canadian Forest Service, flew over the park in 1994 and spied a 75,000-acre patch of "red glow" by fluke. He was on his way to scout for spruce beetles in a place called Rainy Hollow near the Alaska–British Columbia border when a break in the mountains revealed the crimson mark of the beetle. Garbutt suspects the outbreak began in 1990. It seemed to peak in 1997. The outbreak, however, never really quit: to this day the beetles are boring away on the northern and eastern fringes of Yukon's previously infected core. "I've never known of a spruce beetle infestation that lasted twenty years. That's just unprecedented," Garbutt says. To date, the beetle has killed approximately 100 million trees, more than half the mature trees in the region. Government officials, sounding like Alaskans, call it "the largest and most intensive beetle outbreak ever recorded in Canada."

Beetle behavior in the Yukon confounded Garbutt the same way it flummoxed Holsten and Berg. Instead of starting in river valleys and moving up mountainsides, the beetles attacked dry alpine trees and then sauntered down. "It was the exact opposite of what the spruce beetle normally does." In 1995, a winter cold snap settled over the Kluane region, lasting a week. Although the cold air pooled in valleys, killing lots of beetles (Garbutt even found dead adults in the roots of trees), it didn't touch bugs nestled under the snow at higher and significantly warmer elevations. "The next year the outbreak continued on without a whimper."

Garbutt also noted that the beetles concentrated their attacks at the base of trees, for maximum killing power. "They'd only go up as high as five feet and peter out, with the result that it actually took fewer beetles to girdle the tree." The technician had never seen that before. After a while he stopped making predictions based on the records of previous outbreaks. "Every time I opened my mouth, I was proved wrong."

After two consecutive wet and cool summers, Garbutt thought the outbreak would slow down in 2004. But that summer proved to be the hottest on record, and the beetle infestation tripled in size to 245,000 acres. *Dendroctonus* even rebounded in some areas. After skipping stands of fifty- to sixty-year-old trees in search of older wood, they returned four or five years later to take down the adolescents. Again, said Garbutt, "I'd never heard of that before." The beetles attacked even crooked and deformed alpine spruce less than four inches in diameter and twenty-five feet in height.

As millions of white spruce trees grayed and died, the red squirrels panicked. The talkative creatures get most of their nutrition from the protein-rich seeds in spruce cones. But by 2002, seed production had declined so precipitously that the squirrels started to eat beetles. Throughout the outbreak zone, squirrels frantically ripped bark off the trees in a desperate hunt for beetle larvae. For a couple of years, beetle grubs actually made up 20 percent of a local squirrel's energy needs. But it was a short-term solution to an impending famine.

Throughout the Yukon, Garbutt saw lots of weird insect activity caused by tree-shattering drought. The balsam bark beetle, for example, invaded high-country fir forests in the Laird drainage east of Whitehorse, where, Garbutt says, "It's still killing vast forests." The aspen leaf miner, a determined leaf chewer, also became a fixture of Yukon aspen forests, as it did throughout Alaska. "Nobody paid any attention to it. Now it's everywhere." He also saw spruce and pine trees between Whitehorse and the Kluane perishing from outright dryness. "Drought just set the table for the beetle."

The Kluane eruption piqued Ed Berg's attention, too. In 2001, Canadian park ecologist David Henry invited Berg to do another tree ring study to determine what was going on. The two ecologists drilled cores from 433 white spruce trees located in four different areas around the park. Henry thought for sure they'd find that the beetle had visited the historic park on a regular basis. But the tree rings told a different story. Other than some sparse activity between 1934 and 1942, growth rates had remained stable as far back as the Little Ice Age in the 1700s. Until a series of intensely warm summers occurred in the 1990s, bitter cold and regular fire activity had confined the spruce beetle to rare and short-lived appearances. Berg concluded that "widespread, severe spruce bark beetle attacks are rare events in the Kluane Forests"—a twentieth-century phenomenon. In Alaska, climate change aggravated a natural event; in the Yukon, it allowed an invasion of largely virgin terrain.

Like the Alaskans, Yukoners engaged in testy debates about what to do with all the dead wood that, as one local put it, was now just sitting there "waiting for lightning to fire it up." Some Yukon politicians declared the beetle an accomplished terrorist that threatened the health of the region's forests and natural resources. Premier Dennis Fentie even argued that the park "should be declared a disaster zone." Only intense logging and government management could mitigate the wave of dying trees, Fentie argued. When a Canadian journalist later phoned up Ed Berg to ask if any of the government rhetoric was accurate, the ecologist laughed. "You just don't stop those kinds of large-scale outbreaks by cutting down trees. It would be like trying to stop the tide by dipping water out of the ocean." In 2008, a very somber *State of the Park Report* concluded that the melting rates for Kluane's glaciers had tripled and that its forests were now "undergoing abrupt change" because of bark beetles. The report gave both the glaciers and the trees "yellow status and a downward arrow." Changes in the climate, it added, "are contributing to a deterioration of the park's ecological integrity."

AFTER THE great beetle storm, life changed in the spruce forests of southern Alaska. Before the attack, two tons of woody debris per acre littered the forest floor. After the beetle, scientists measured forty tons per acre. The Kenai is now the house that *Dendroctonus* built. Canada bluejoint grass took advantage of all the sunlight and exploded, smothering everything in its path, including spruce seedlings. The six-foot-high grass not only chokes out other plants but lowers soil temperatures and makes a spectacular early-summer fire hazard. In 2007, authorities spent $8 million to control one such escaped grass fire in the Caribou Hills. Most foresters now predict that the spread of bluejoint grass may delay tree regeneration on the peninsula by forty to one hundred years. Others think parts of the peninsula could end up looking like the grasslands of southern Alberta.

Different animals now use the changing forest in different ways. Birds that don't nest in trees, such as the hermit thrush and the blackpoll warbler, are thriving. After years of rich pickings, woodpeckers are now contemplating poverty. Moose numbers have declined since the 1960s and show no signs of increasing; there is not much to browse on in a beetle-killed forest. The red squirrel simply lost its home. With fewer trees sucking water from the ground, stream levels have risen, with unknown impacts on salmon. The snow melts faster in the spring because the great shade trees are gone.

It's hard to find a spruce beetle in south Alaska these days, yet the insidious footprint of climate change keeps growing bolder. On average, the winters continue to warm three times faster than the summers. Wetlands that used to be fuel breaks have become fuel bridges as they fill in with grass and black spruce. In Homer, apple trees now grow fabulously well. Three years ago the city did something few U.S. cities have done: it cobbled together a Climate Action Plan, because "Alaska was ground zero for global warming." In the meantime, the Kenai Peninsula grows less wild by the day, thanks to a surging wave of fifty thousand scenery chewers and southern urban refugees. Every year, cars, planes, and cruise ships disgorge a million tourists to fish in the fjords or gawk at the mountains. In Homer, the "Cosmic Hamlet by the Sea," the local halibut limit remains two per person.

By 2002, after he'd documented the climate trigger and more than half the forest in the wildlife refuge had been lost to beetles, Ed Berg wondered what the Kenai might look like forty years on. So he took a trip in a Cessna 185 to an old-growth spruce forest felled by the beetle in 1958 at a place called Turnagain Arm. Photographs taken in 1975 showed mostly bare trees still sticking up like toothpicks. In 2002, Berg found the beetle kill prone on the ground and covered with moss. Here and there, among surviving spruce, stood lots of birch trees. Berg breathed in the "moist feel of the place." Amid a tangle of rotten logs, he also discovered various sizes of spruce seedlings, flourishing under bushes and devil's club. Many of the seedlings were growing on rotten stumps, or nurse logs, germinating in the air in the way that old-growth forests renew themselves. Berg couldn't run through the new-growth forest, as he likes to do, because of the logs on the ground and the thorny arms of the devil's club. But the bucolic scene reminded him that life goes on after a trauma.

Berg used to walk through the Kenai's skeletal forests and say, "Egad, look at all those dead trees." Now he marvels at the impressive new growth, already six feet high with foot-long leaders in some places. But he doesn't think the new trees will grow very old. All the climate forecasts show sustained off-the-chart warming, which means a new crop of spruce beetles will harvest the new growth as soon as it reaches twelve inches in diameter. It's unlikely that Alaskans will ever again see old spruce giants with diameters of two to three feet on the Kenai, adds Berg. Leafy birch, alder, and aspen will replace the gentle conifer giants. "The long and short of all this is that now is a bad time in history to be a conifer tree of any kind." Alaska, Berg cautions, "is a warning of things to come throughout North America."

Northern Europe can also expect dramatic changes. Furious windstorms have upended spruce forest plantations in Germany, Sweden, and Switzerland, prompting unprecedented outbreaks of bark beetles. In 2008, a Viennese study concluded that climate change could increase bark beetle timber losses by 219 percent. Researchers at the Norwegian Forest and Landscape Institute suggest that the climate question for the beetle has really become a Shakespearean one: "To be or twice to be?" *Ips typographus* normally reproduces just once a year in Scandinavia but twice or even three times a year in warmer parts of Europe. The forests around Oslo rarely, if ever, witness two beetle flights in a summer. But thanks to climate change, the spruce beetle's ability to mature faster, much like it did on the Kenai, will shift its distribution 400 miles by 2070 to encompass most of Norway. According to researchers, "With two generations per year, there will also be two attack periods on spruce annually, one in the spring and one in July and August." Trees produce less resin in late summer. Combined with more windstorms and drought, the result of a two-generation beetle could be either shorter and more frequent outbreaks or "an increased volume of destroyed forest." In any case, researchers concluded, the shift could have "profound effects" on forestry. It could also change the whole Norwegian experience.

At elementary schools on the Kenai Peninsula, teachers now have students "role-play a tree." The group that plays the bark forms a circle around taller students who pretend to be the heartwood and the phloem, the inner tissue near the bark. The dozen or so bark actors all face out of the circle. The teacher tells them to protect the tree by looking like menacing defensive linebackers. "Ask students if they hear a distant whine and tell them it is a spruce bark beetle about to attack the tree," reads the teachers' instruction manual. Soon one or two students with fingers formed in the shape of drills appear on the scene and abruptly attempt to break through the bark's defenses. According to the teachers' manual, "The bark should try to protect the inner part of the tree."

Nearly a decade after the infestation, Ed Berg and his wife moved away from their denuded East End Road property. Without the giant spruce trees sheltering their land, the wind came up from the bay and chilled the couple. "We had to wear long pants and sweaters in the summer." The Bergs moved into a large house in Homer on a property with no spruce trees but a "spectacular view" of Kachemak Bay. For several years, Sara, a nurse, offered assisted living for the old and terminally ill. "They all had good deaths with their families. It puts your life into perspective," Berg says. He sees no comparison between his wife's work and the ecological services that beetles offer an aging, drought-stressed forest. "Going out with the bark beetle is like going out in a war. I'd rather die gradually of heart rot."

Hardly anyone talks about the beetles in Alaska now. The grief and emotion have passed. But nothing seems quite the same after the extreme killing of the trees. At least that's how it seems to Mary Jane Shows, one of Berg's former neighbors on East End Road. During the summer, Shows, a jazz singer, used to walk out to an opening in her five-acre parcel of land and sing to her grand spruce trees around eleven o'clock each evening. Shows says singing is her medicine. To the forest and her attentive neighbors, she sang "Willow Weep for Me" or "Stormy Weather." "The trees just popped my voice back to me," she says. But then the clouds of beetles descended, and the trees died. "I cried and cried when they cut them down." Shows later moved to Juneau. The spruce beetle, she says, was "our fire. It will never go back to what it was."

The Beetle, the Bus, and the Carbon Castle

. . . . .

"There's a Legion that never was 'listed,
That carries no colours or crest,
But, split in a thousand detachments,
Is breaking the road for the rest."
RUDYARD KIPLING, "The Lost Legion"

STEPHEN L. WOOD extracted his first bark beetle—what Alaskan high schools now call "the unseen villain"—from a conifer with a penknife in Lehman Creek Canyon, Nevada, in 1939 at the age of 14. "The attraction was immediate and permanent," Wood later wrote. When he died in 2009 at the age of 84, the former entomologist at Brigham Young University, Utah, left behind a collection of 80,000 bark beetles carefully pinned to the bases of 181 specimen drawers for the Smithsonian's National Museum of Natural History. (Beetle experts tend to be a bit obsessive.) Wood braved cannibals, leeches, jaguars, and wild elephants to collect them. He became such an authority on the insect that 3,000 entomologists around the world routinely called on him to identify mysterious or unknown bark beetles. After careful inspection of a beetle's many knobs and protuberances, he'd inform authorities whether the specimen was a casual visitor, a real tree killer, or some foreign vandal. Wood probably saved the forest industry millions of dollars.

The undercelebrated Wood also wrote the acknowledged bible on the bug. He conceived the project in 1952, started writing in 1972, and published the work in 1982. The 1,359-page tome, *The Bark and Ambrosia Beetles of North and Central America*, caused his wife and children to endure "decades of hardship and sacrifice," Wood said in his acknowledgments. Before his death, he went on to pen another opus, 900 pages on bark beetles in South America.

Wood's beetle bible is an incredible, though obtuse, work full of maps, photographs, odd anecdotes, and elaborate illustrations of the alien-like parts of bark beetles. It weighs five pounds. Wood had a lot to say about bark beetles in prose as dry as a dead tree. As a taxonomist (a person who names and identifies things), he duly noted that 6,000 species of Scolytidae worldwide (it's now 7,500) belonged to 25 tribes with mafia-like titles such as Tomicini and Ipini. Wood estimated that insects caused 90 percent of all tree deaths and that his beloved scolytids (also called scolytines, due to their genetic kinship to weevils) accounted for more than half of the killing. In fact, he noted, bark beetles brought down more trees every year than fire, disease, and wind combined. The majority of bark beetles bred in "unthrifty, broken, over mature and dying woody plants." But a dozen or so species aggressively attacked healthy trees. The majority of these economic troublemakers lived in northern pine and spruce forests, Wood wrote, and belonged to the clan or genus *Dendroctonus*, which in Latin means "tree killer." A couple of beetle hit men also belonged to the prominent *Ips* tribe.

Nevertheless, Wood admired the bark beetle as an impartial manager of aging trees and forests. Bark beetles, he wrote, eliminated mature or stagnant trees and recycled them as forest nutrients. When trees crowded each other or got sick or injured, beetles sorted out the winners and losers. By disposing of unthrifty trees, the insect also opened "the way to more vigorous growth of the surviving plants." It was obvious to Wood that when God made the bark beetle, he really didn't make a pest. But "because their natural function in a managed forest conflicts with human interests," professionals invariably transformed the beetle from a "beneficial to a destructive element." Wood also argued that the human population boom in the twentieth century created such insatiable demands for wood products that government, industry, and landowners lost their patience for beetle forest management and renewal. Long ago, when the beetle ruled the woods, "it wasn't particularly important that several hundred years might be required to restore the original forest." Now that mattered. As a consequence, scolytids required extraordinary control efforts as well as thick books explaining their identity and character.

Wood highlighted the beetle's many remarkable qualities. Hardly any insect can make a living inside the dark and inhospitable environment of a living tree except the aptly named bark beetle. As with all empires, scolytids send out hardy and brave "pioneers" to scout for new homes before the real colonists arrive. Climate and drought resolutely shape the beetle's behavior and success. During boom years, bark beetles can launch massive attacks in biblical swarms. In the tropics, Wood often watched clouds of ambrosia beetles, which breed in fresh dead wood, circle a tree about to be cut and then riddle it with holes before it could get to the mill. It's instructive that only humans, wolves, hyenas, lions, killer whales, piranhas, and ants also do what communal bark beetles do: hunt larger prey in social packs. (Depending on the vigor of lodgepole pine, an attacking beetle swarm can inflict anywhere from 58 to 112 bullet-like entries per square yard.)

Wood also noted that swarming bark beetles talk to each other by brewing a unique blend of chemical perfumes that can attract or repel fellow attackers as well as potential mates. Wood thought these pheromones "played a vital role in the survival of the species." He also described the beetle as a "weak flier" that took advantage of wind currents to move long distances. He asserted that flooding could transport the beetle a long way, too. Wood once kept living larvae "immersed in water cooled to near freezing point" for two hundred days to prove that a flood-swept log could produce a living beetle after floating hundreds of miles.

Wood offered comments about bark beetles' dining habits as well. "Some species confine their attacks to the cones of their host, others infest only tiny twigs, others small branches, limbs, boles or roots. Some breed only in shaded-out branches of standing living trees, others in felled or broken material." The red turpentine beetle, for example, favors ponderosa pine and rarely kills the tree. The pine engraver goes after drought-stressed lodgepole and ponderosa pine. The western pine beetle targets overgrown mature "decadent" ponderosa pine. The fir engraver chews on the branches and tops of grand fir. And the relationships are very particular. The mountain pine beetle farms and then destroys entire forests of lodgepole pine. The Douglas-fir beetle and the spruce beetle reproduce best in wind-felled trees. The southern pine beetle has a hankering for pines struck by lightning. Another beetle dines on the berry of the coffee bush. And so on.

The beetle's ability to engrave elaborate winding galleries under the bark of a tree also intrigued Wood. Whereas ambrosia beetles make simple caves for growing fungal gardens, bark beetles mine long tunnels shaped like stars or the letters J and S. An entrance tunnel leads to a nuptial chamber where adult beetles can have sex in private. The female then carves out numerous straight lines in which to lay as many as three hundred eggs. The larvae branch out in different directions in their own tunnels.

Some galleries can be a foot long and look as complicated and extensive as a crazy suburban road network. The fir engraver creates a startling gallery that resembles a Christmas tree. Given the wee size of the carvers, Wood called the scale of the galleries "enormous."

Not much has been written about the sex lives of bark beetles, but Wood thought the insect displayed "unique reproductive behavior." Most *Dendroctonus* beetles practice monogamy, but some *Ips* species (like some Mormon clans) create remarkable harems composed of several females. The monogamists typically court like young teens, with stinky perfumes and playful touching, while the males make loud clicking sounds. After chewing through a tree's bark, a female sends out an alluring scent to draw a male to the entrance of her tunnel. The suitor performs a mating song by scratching one of his body parts. This stridulation catches the female's attention, and the two then court and mate. After sex, the male seeks out other partners or hangs around to keep house. For some species, that includes keeping the galleries free of frass, chewed-up wood mixed with beetle shit.

Wood devoted the bulk of his book to describing the 1,430 species of bark beetles in North and Central America. He patiently told readers how to spot the difference between bark beetles such as *Monarthrum infradentatum* Wood and *Monarthrum corculum* Wood. It has something to do with "elytral declivity."

ASK ANY expert about the bark beetle, and every single person will identify something queer or just plain wonderful. John Borden, a Canadian chemical ecologist, admires the bug because bark beetles bury their dead and are smarter than humans. "Every time I think I've gotten to understanding how they work, they blindside me and do something different," he marvels. Barbara Bentz, a Utah entomologist, can't believe the bug's damnable fragility: "You can't even raise them in the lab. I'll take them off bark and put them in my hand and they'll crawl up and wobble and open their wings and fall over. It's amazing how clumsy and successful they are." Richard Hofstetter, a researcher at the University of North Arizona, finds the insect a study in other-worldliness. "It lives one part of its life in the dark chambers of a tree and the other part as a free-flying organism. And it's either a bust or boom mode," he says. Diana Six, an entomologist and pathologist at the University of Montana, regards the bark beetle's diverse partnership with fungi as dumbfounding. "The beetles have parlayed their relationship with fungi into becoming the most successful insect on the planet. If they didn't have these fungi, they wouldn't be known to us." Ken Raffa, an ecologist at the University of Wisconsin, cites the beetle's ability to cross biological borders in unfathomable numbers. "They can go from living on such a small level to operating on such a big scale that they can change entire ecosystems, if not carbon cycles." Raffa also thinks the bark beetle keeps us honest. "We don't want to make decisions in forests on a scale of one hundred years, but they force us to grapple with our own special interests," he says.

All in all, the bark beetle story weirdly combines science fiction with comedy and chaos theory. It's striking that a bug that can undo a forest can also be squished by a thumb. A bark beetle may look as sleek as a small-caliber bullet, but it flies as awkwardly as a puffin. The head profile of a mountain pine beetle resembles that of Darth Vader; the spruce bark beetle could be mistaken for the tip of a ballpoint pen. Both beetles behave like total wimps. Once flipped onto their backsides, the insects can barely right themselves. Nevertheless, a swarm of pine beetles can take down a thousand-year-old white bark pine in a few hours or decimate a mature lodgepole forest larger than Switzerland in a couple of years. Given that trees have developed defenses as complicated as well-fortified medieval castles, the average beetle does more dangerous work than a U.S. soldier in Afghanistan.

Yet small creatures can't defeat giants without a community of determined associates, and here the beetle story becomes a trek into a miniature world that is still largely terra incognito. No beetle forest pioneer or legionnaire works alone. The sleekly cylindrical exoskeleton of a bark beetle serves as a kind of psychedelic bus crammed full of combative and bloody-minded helpers. And everyone, it seems, has come prepared for the most lethal forms of chemical and biological warfare.

The Bus

Although Stephen L. Wood knew that the bark beetle carried fungi and mites, the old man had no idea how crowded the flying vehicle was. Or how important some of the passengers—an astounding array of fungi, mites, yeasts, viruses, protozoa, nematodes, and bacteria—might prove to be. Each one could command a Wood-sized tome. In 1927, the great animal ecologist Charles Elton, while studying the curious rise and fall of mouse plagues, took a moment to count all the passengers on a field mouse near his lab at Oxford University. He found a wild bunch of characters: "The ear had a tick larva and some mites; the fur had at least a dozen sorts of mites, a beetle, eleven kinds of flea, and a louse; the skin had most of these as blood-suckers; and also a more persistently attached adult tick and a kind of mite causing scabs on the limbs; the anus and genital organs had the harvest mite; the liver had a tapeworm larva whose adult comes in cats... The small intestine was the abode of three other roundworms, two tapeworms, three flatworms and six Protozoa..." All in all, the mouse carried at least forty-one species.

The zoo that rides a bark beetle is equally diverse and surprising. The spruce bark beetle (*Dendroctonus rufipennis*), for example, can carry up to 10 different species of fungi, 6 different kinds of mites, and 9 species of bacteria. Under its wings, the beetle also transports a species of nematodes, or worms, capable in turn of transporting 4 different types of fungi. From 10 different species of pine and fir trees in Mexico and Guatemala, researchers have isolated more than 403 strains of yeast in the guts, ovaries, eggs, and shit of beetles. The southern pine beetle can carry up to 100 different species of mites, plus bacteria and scores of fungi. Every day scientists discover more and more passengers on the bus.

Diana Six, a forest entomologist and a fungal expert, suspects that the most important riders in this crazy vehicle are the fungi and the yeasts. The association between beetle and fungi is so ancient and nuanced, she explains, that many bark beetles even have special suitcases (mycangia) in their jaws to carry fungal spores from tree to tree, much the same way a car has a glove compartment. The beetles also have convenient pits on their exoskeleton for transporting fungi.

After Stephen Wood died, some functionaries at Brigham Young University proposed to discard leftover copies of his life-work on bark beetles. When Six heard about the debacle, she intervened. A stack of 30 copies now sits in a corner of her lab for students. Beetle researchers call Six "the fungal guru," and for good reason. The athletic body builder and researcher started collecting insects and fungi as a young girl. After various diversions, including a stint as a motorcycle biker and drug addict, she began studying the interaction between fungi and beetles and has now been doing so for nearly 20 years. Having lived on mean streets of Los Angeles, Six knows how complex, fragile, and surprising the world can be. She keeps about 1,200 samples of fungi, representing nearly 100 species from 15 different beetles, in her refrigerator at the University of Montana in Missoula. Some of the molds look light and golden brown while others are dark and granular. She has even made a fine beer from yeasts carried by bark beetles. Using a special technique, she harvested spores from a beetle's mycangia and then grew the yeasts on agar plates.

"One of my students was a brewer," she explains. "So we ran fermentation tests for sugars, and a few were really good." She branded the potent beers "Six Legged Ales." She named the light brews "Callow" after young bark beetle adults and called the darker ale a "Porter." The beer sells quickly at the annual Western Forest Insect Work Conference.

Most of the fungi carried by bark beetles belong to Ascomycota, a huge and ungainly phylum that has produced the majority of the world's antibiotics. "Ascomycetes are to fungi what the beetle is to the animal kingdom," explains Six. The molds outdate beetles altogether, and they behave outrageously. Fungi tend to be fiercely competitive. They reproduce either sexually or asexually and usually feed on the outside of their bodies. Like unhappy vampires, they have the ability to live forever, and they grow and grow. "They are incredibly effective at invading and gathering nutrients and transferring those nutrients to their bodies," says Six.

Fungi and beetles often work together, like an old farming couple. Some fungi provide beetle nutrition. Others ward off unfriendly fungi. Many are just looking for a free ride to exploit a wounded tree. Just about every member of the *Dendroctonus* clan has two to three established fungal associates, and new associations keep popping up every day, thanks to molecular technology. "The number of new species being discovered is unbelievable," says Six, and range in the hundreds.

Beetle larvae and young adults depend on crops of fungi, because bark and wood chips make a poor diet. Some beetle adults can't reproduce without dining on fungal spores before they leave a tree. The nutritious fungi carried by the mountain pine beetle also offers the beetle climate insurance, by being either cold- or warm-tolerant. "Beetles are smarter than us. They always have a backup fungus. If they just had one and conditions changed, they'd be screwed," says Six. Basically, the fungi allow the beetle to use a marginal habitat and do it well. "Without fungi, the beetles would be nothing. It's the most amazing thing."

Several years ago, Six performed an interesting experiment. She studied mountain pine beetles as they attacked three different types of pine logs. One set of logs contained a very good fungus, another had a moderate fungal meal, and the third had no fungi at all. The beetles with the benevolent associate produced bigger broods of fitter beetles. Strong, better-nourished beetles live longer and fly farther. The beetles with the casual passenger produced okay offspring, but nothing special. And the beetles without fungi did damn poorly. "They were tunneling and tunneling in the phloem," Six says, "as though they were looking for something. They also created long feeding galleries, which indicated they were trying hard to get food."

Given the fungi's importance as a food source, Six believes that they probably play a key role in beetle population explosions. The sugar-rich inner bark of a tree contains little nitrogen, an essential nutrient for herbivores. Fungi, however, can tap into nitrogen in a tree's sapwood and increase the amount of nitrogen available for beetle larvae in the inner bark by 40 percent. "That's astronomical for a herbivore," says Six. "It's like eating junk food and bringing along a nutritional supplement. The beetle couldn't make a go of it without the fungi." Beetles carrying fungal spores resemble farmers with bags of seeds: they both have crops to sow. A beetle explosion in a forest, a primitive experiment in fungal farming, looks a lot like an infant civilization that has just discovered wheat or corn. Six also suspects the fungi provide components beetles need to make hormones in huge quantities essential for beetle molting and egg laying. "Nobody has ever thought much about the fungi or asked what's their agenda. Everything has focused on the beetle."

Other passengers of great importance are hordes of mites. These small, headless creatures, which look like weird spiders, take up a lot of room on the bark beetle bus. Mites (ticks belong to this family, too) are so small that a million can thrive in just a square yard of forest. One acarologist wrote several years ago that mites dwell "in a strange and beautiful world where a meter amounts to a mile and yesterday was years ago." Overall, mites rival beetles in their abundance and diversity.

A typical bark beetle can ferry up to 50 mites, representing as many as 10 different species. (About 40,000 species of mites have been identified, but scientists suspect there are millions.) "Mites don't have wings. So it's walk or take an airplane," explains John Moser. The eighty-one-year-old Louisiana-based acarologist fondly describes the creatures as "little arthropods with four sets of legs instead of three." Moser, a contemporary of the ant expert E.O. Wilson, has been studying mites and beetles for more than 50 years. Although the eminent specialist has officially retired, he remains the only mite expert employed by the U.S. Forest Service. Moser suspects that 99 percent of the world's bark beetles carry mites, "but hardly anyone knows what they are really up to." Some eat fungi, some dine on beetle eggs, and others prey on nematodes. A whole bunch just travel with fungi in their guts.

Bark beetles and mites can be a complicated power couple. Moser recently looked at the two principal mites transported by elm bark beetles that carry the notorious Dutch elm disease. (A Dutch pathologist first identified the Asian fungus that clogs the water transport of elm trees and then wilts the leaves and branches.) The nasty fungus, which appeared in Europe around 1918 and migrated to North America in 1930, has killed millions of elm trees. It has transformed traditions, societies, and landscapes on two continents.

Given that it takes about a thousand fungal spores to wilt an elm tree, scientists originally assumed that the beetles carried the fungus in pits on their skeletons. But Moser found that two mites alone could carry enough sticky spores to destroy a tree's health. "We counted as many as two thousand to three thousand spores in the guts of one mite. That's all it eats." Because the fungus makes the phloem inedible, it's really not in the interest of the beetle to infect the tree. But it *is* in the interest of the mite, which dines on the stuff. Moser suspects that Dutch elm disease is driven more by the mites on the bus than by the bus driver, the bark beetle.

In addition to fungi and mites, beetles also carry a wild assortment of bacteria. Nobody knows how many bacteria ride the bus, but scientists have found them in beetle eggs, galleries, mycangia, and even beetle shit. The spruce beetle secretes bacteria on the floors of its galleries to quell the growth of competing fungi. Probably every member of *Dendroctonus* packs its own emergency bacterial medical kit for either promoting or thwarting fungal growth in its egg galleries.

Consider for a moment the strange existence of the mite-infested southern pine beetle. Richard Hofstetter of the University of Northern Arizona has studied this crazy biological bus for several years and says scientists are still kicking at the tires.

Unlike other members of the *Dendroctonus* clan, which blanket a forest, the southern pine beetle attacks longleaf and loblolly pines in spots and patches. It starts with a few trees, moves on to ten or more, and then assaults up to fifty by snaking a path through the woods. "It's like a fire," says Hofstetter, a small, quick-witted man. Watching the southern pine beetle in attack mode sounds a bit like watching an invisible monster paint a street-wide trail of yellow and red through the forest. A big infestation often starts on a smoldering tree struck by lightning. "They are magnets for the beetle," Hofstetter explains.

After drilling a hole through the bark of a mature loblolly pine, female beetles typically plant one or two friendly fungi under the bark. These molds soften up the tree and provide a secure source of nitrogen for the larvae. As Hofstetter notes, the pine beetle attacks "living trees, because it's the only milieu where these fungi can survive, and the fungi allow [the beetles] to explode in the numbers they do." As the beetles attack more trees and start a feeding frenzy, they pick up more and more mites. These mites, which hang out on trees, pretty much depend on a beetle population explosion to enjoy expansive living. During an outbreak, it's easier for them to find a host. As beetle numbers soar, so too do mite populations. "As many as twenty to one hundred mites will jump on the beetle," says Hofstetter.

But the tiny hitchhikers make problematic passengers, because they carry spores of an antagonistic blue-stain fungus, *Ophiostoma minus*. This fungus makes a great snack for mites, but a toxic meal for beetles. Once *O. minus* starts invading a tree's phloem, nearby tunneling larvae will try to move away from the invader. The bully-like fungus is also mildly pathogenic to the pine tree. So a beetle jam-packed with mites carrying *O. minus* carries the seed of its own destruction. As the blue-stainer multiplies, it crowds out all the nourishing fungi. In fact, population booms of mites carrying the fungus have probably stopped some southern pine beetle outbreaks.

But in 2008 scientists discovered another beetle novelty. In its fungal suitcase, the southern pine beetle often also carries bacteria that are able to zap and neutralize the mite-borne antagonistic fungi. There aren't many medications that can obliterate deadly fungi infections in humans, but in the empire of the beetle just about anything seems possible. To defend its offspring, the southern pine beetle handily deploys one of the world's most powerful antibiotics. How other bark beetles deploy bacteria remains a mystery.

The Castle

For hundreds of years, peasants, loggers, and beetle fanatics have wondered how an insect as diminutive as a grain of rice could overwhelm something as formidable and long-living as a tree, let alone millions of trees. "It seems hard to imagine that such a small insect could kill a large tree," wrote the German botanist Johann Friedrich Gmelin in 1787. But Gmelin offered a good answer: "If one considers how extraordinarily fruitful the beetle is, and attacks the tree not individually, but in entire groups of thousands, then it doesn't seem impossible that it can overpower a tree." He wasn't far wrong.

Swarming a thirty-foot-high conifer as thick as a concrete block isn't much different from besieging a medieval castle. It takes numbers, brute force, good communication, and an element of luck. The ancient and frenzied contest pits a large, stationary, aging protagonist against a small, highly mobile predator. In this hazardous encounter, both sides deploy their weapons with abandon. A well-defended tree fiercely protects its keep with thick bark walls, moats of deadly gases, and cauldrons of sticky resins. The beetles arrive as thousands of cylindrical buses full of well-armed passengers. As the bugs direct and time their assault with chemical signals, all hell breaks loose. A frenzied swarm of beetles can bring down an old tree the same way Viet Cong guerrillas crippled the U.S. industrial war machine or the Taliban defeated the Russian army. It's all about the power of small things and the impact of the highly improbable.

Trees can marshal formidable defenses, among them copious amounts of resin full of toxic hydrocarbons. Some fir trees sport bubbles of the stuff on their bark, and most pine carry the toxic goo in their ducts. Like a cauldron of boiling pitch, resin can smother an attacker or gum up an insect's working parts, including the powerful beetle jaws that move sideways like a pair of scissors. Resin can also gas nearby beetles with noxious vapors while at the same time sealing and repairing bark wounds. The fearful effectiveness of resin explains why so many bark beetles prefer attacking windblown or lightning-struck trees, which can't muster much of a defense. Resin also explains why bark beetles

carefully pick the weakest tree. Beetles
have learned that it's easier to overcome an
aging, drought-stricken king than a knight
in shining armor.

Resin makes an amazing weapon. It consists of thirty or more chemical terpenes, or
what ordinary folks know as turpentine.
These powerful plant-made hydrocarbons,
which can float on the air, smell like old-
fashioned paint varnish. Until oil-based
paints came along, the resin from conifers
made the stinkiest and most widely used
paints and thinners in the world. Twenty-

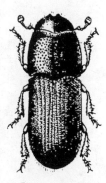

SOUTHERN PINE BEETLE
(*Dendroctonus frontalis*)

five percent of all "aroma chemicals" sold today are still made
from turpentine. The terpenes found in tree resin can repel
not only beetles but most of the passengers on the beetle bus,
including pathogenic blue-stain fungi. Conifers have been triumphantly entombing bark beetles in resin for hundreds of millions
of years. Golden drops of amber containing defeated bark beetles
and their mites have been found in the remains of ancient forests
in the Dominican Republic, the Baltic, Lebanon, and Myanmar.
Probably more bark beetles have been found preserved in 25- to
40-million-year-old Dominican amber than any other insect.
The beetle pitch wars are so ancient that scientists discovered
an engraving by a successful sapper on a piece of 45-million-
year-old mummified wood from the high Arctic in 2009. But
while healthy trees can produce large amounts of resin, drought-
stressed trees can't. And a tree without resin is a castle without
boiling oil.

The siege of a lodgepole castle formally begins when a female
mountain pine beetle, the pioneering sex, selects a vulnerable
tree. Nobody knows exactly how a female killer picks her vic-
tim. But if she selects poorly and hits a duct of resin, she will get

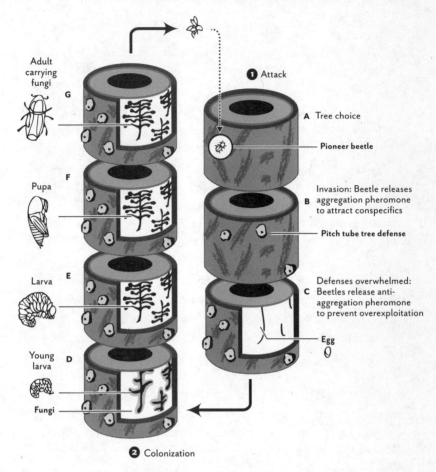

Adult carrying fungi

G

Pupa

F

Larva

E

Young larva

D

Fungi

❷ Colonization

❶ Attack

A Tree choice

Pioneer beetle

B Invasion: Beetle releases aggregation pheromone to attract conspecifics

Pitch tube tree defense

C Defenses overwhelmed: Beetles release anti-aggregation pheromone to prevent overexploitation

Egg

"pitched out" and experience "death by engooment." A successful attacker, however, quickly chews her way through the bark into the phloem. In the process, she'll inhale the tree's chemical resins and with the help of yeasts turn them into a powerful scent (trans-verbenol). It attracts more siege-mates, but mostly males. (The remarkable mid-gut of a bark beetle not only digests wood but serves as a chemical factory.)

Male attackers come equipped with their own chemical arsenal. When young males leave their respective tree nurseries and take flight, they swallow air, forming a large air bubble (an internal barometer) in their mid-gut. While in flight, they convert body chemicals into another pheromone, exo-brevicomin. By the time they reach a female pioneer, the flatulent swarmers are ready to fart out this scent. In turn, their redolent wind attracts even more beetles, but this time mostly females. During the siege, flying beetles of both sexes pick up the mass-attack perfume and swarm the castle like Mexican regulars at the Alamo. "The more the tree tries to fight, the more the beetles pour into the tree. It's an early version of jiu jitsu," explains beetle expert Ken Raffa.

Depending on a tree's vigor, it may take anywhere from ten to ten thousand beetles to seize it. It all depends on the health of the castle's defenses. About six hundred mountain pine beetles can take down a mature lodgepole, while more than two thousand southern pine beetles are needed to dispatch a loblolly pine. To regulate the attack density, beetles exude several different kinds of repellent "perfumes." The most potent, verbenone, is cleverly converted by the beetles and their yeasts in one simple step from the attractive pheromone trans-verbenol. Without these potent repellents, "you'd have an underpopulated slum, or Hong Kong plus," explains chemical ecologist John Borden. "The chemicals help them space out in the tree quite well, so each family has a good place to raise the kids, and then they create a gated community to keep out the homeless."

As the attack intensifies, a tree will ramp up the toxicity of its chemical defenses a hundredfold. "That's why attacking a tree is an extremely dangerous thing to do," says Raffa. The rich turpentine aromas also send out help signals from the tree to various beetle predators. At this point, wasps and clerid beetles will come to the tree's assistance, even picking off attackers as they begin to tunnel in the bark.

Once the bark has been breached, the tree employs a second line of defense. It tries to isolate beetles or fungal growth in the phloem by encircling them with deadly toxins that kill all surrounding tree tissue. In the process, hundreds of beetles may be gassed or sealed in a tree. Yet a pine wounded by thousands of beetles quickly runs out of resin, energy, and toxins. "If you were attacked by ten guys in a bar, the fact that you could manufacture some resistance almost becomes irrelevant," adds Raffa. As more beetles bore into the bark, the tree transforms its metabolism entirely. It stops producing sugars and concentrates on pumping proteins in a last-ditch attempt to deprive the invaders of nutrition as they tunnel their galleries in the inner bark. Most sieges last but a day or two. As soon as the beetles realize the tree has reached the point of no return, they secrete verbenone, another perfume, en masse. This scent tells other beetles that the castle has been stormed and the keep is now full. The bark beetle expends only the energy it needs to kill a tree, and no more.

For years, many beetle experts blamed tree death on the pathogenic blue-stain fungi carried by most aggressive tree killers. They assumed that the fungi served as the battering ram to break the tree's defenses, because they found the stain in most beetle-killed trees. It was believed that the blue-stainers fatally clogged the tree's plumbing. But Diana Six doesn't think the story is that simple. Nothing is, in beetle country.

For starters, both the southern pine beetle and the spruce beetle can kill trees just fine with or without blue-stain fungi. When scientists hammered a pine tree full of hundreds of beetle-sized holes (the experiment happened only once), the tree died with no fungal intervention. Moreover, the most aggressive bark beetles and most effective besiegers tend to carry mild fungi for nourishment instead of virulent blue-stain varieties for killing. Finally, blue-stain fungal colonization of a tree's sapwood takes several weeks. By then, beetle larvae are already feeding on the inner bark, and the girdled tree is pretty much defeated. Says Six, "The beetles are girdling and tunneling away. The tree declines before the fungi takes off."

The blue-stain fungi probably do play an essential role in the siege, but not necessarily as tree killers. Six has found that they not only provide key nutrients for insects but help to manage competing colonies of quarrelsome fungi on a dying tree. Many might also be there to take advantage of the breached castle. The deadliness of blue-stainers, rather than being useful in bringing down a tree, may simply ensure fungal survival in a hostile environment. Because blue-stained wood absorbs moisture readily, the fungus assuredly accelerates rot in a dead tree. Like blood on a battlefield, though, the blue-stainer may just mark the scene of a struggle.

Nobody knows for sure if bark beetles taught humans how to murder trees, but the insect probably served as an ancient instructor. From Europe to Central America, early farmers cleared great forests by severing bark with stone axes. Gouging a ring around a tree starved the tree of nutrients; farmers would later burn the woody skeletons, thereby opening up the land. American farmers became so proficient at girdling trees (it was cheaper than clear-cutting) that such forest-destroying practices were known as the "Yankee" method in the seventeenth and eighteenth centuries.

It's important to remember, the experts point out, that when beetles storm castles across a landscape, they are sending out two crucial messages. The first is that change is not gradual. Big systems don't fail slowly or fall apart in straight lines; they unexpectedly crash and burn. The second is that "insects are only what we see in the forest," advises Diana Six. "They are just responding to signals. They are not the cause. They are finishing the trees off. But the trees are being set up for them."

And that's what Allan Carroll found in British Columbia.

THREE

The Lodgepole Tsunami

. . . . .

*"There is no human being who is not directly or
indirectly influenced by animal populations, although
intricate chains of connection often obscure the fact."*

CHARLES ELTON, *Animal Ecology*

ALLAN CARROLL, one of Canada's foremost insect ecologists,
first walked into a beetle-attacked forest at the age of twelve.
He never forgot the sensation. Carroll's father then worked at a
pulp mill on the coast in Gibsons Landing, British Columbia, but
the family preferred to vacation in the hilly meadows of cowboy and Indian country in B.C.'s Chilcotin region. This majestic,
rambling land looks much like the high country of Montana,
Colorado, and Wyoming. During the mid-1970s, the mountain
pine beetle invaded the Chilcotin's aging lodgepoles. The young
Carroll walked through the battleground wondering "what was
killing all the trees."

During that outbreak, the beetle took out about two and a half
million acres of mature pine. Two –40-degree-Fahrenheit winters
abruptly ended the massive feed in 1986. Carroll recalls coming

across red-topped trees and huge trunks marked with scores of sticky pitch tubes. The air was positively thick with turpentine. "You knew something was going on," he says. "I'd peel off the bark of dead trees and see tunnels but no beetles. Just the results." The penetrating and volatile smell of an attacked forest, much like the acrid smell of fear, still lingers in his mind. "You can smell it whenever you are in pine country."

And that odor is exactly what Carroll inhaled when he stepped off a Bell 206 helicopter onto the shores of Eutsuk Lake in British Columbia's Tweedsmuir Provincial Park some twenty years later, in the early summer of 1999. Carroll, a methodical thirty-five-year old, had just spent eight years battling defoliators in spruce and fir forests on the east coast with the Canadian Forest Service. When the Pacific Forestry Centre in Victoria, B.C., offered the west-coast boy a chance to come home in 1997, he grabbed it. Les Safranyik, a Hungarian-born forester and renowned pine beetle expert, was on the verge of retiring, and he offered Carroll the file on *Dendroctonus ponderosae*, the mountain pine beetle. The Canadian Forest Service had been studying the creature for a hundred years, but the field was still pregnant with unanswered questions, Carroll remembers, such as how the insect went so quickly from zero to "up to your arm pits in numbers."

To answer that and other questions, the wardens of Tweedsmuir asked Carroll to check out a growing patch of red-topped lodgepole trees, or "faders." The wilderness park, which covers over 2 million acres of largely inaccessible wilderness in the west-central part of British Columbia, was vulnerable. Mature lodgepole pine dominated much of the spectacular mountain landscape.

To Carroll, who would spend the next decade as Canada's go-to guy on the pine beetle, the Tweedsmuir outbreak seemed curiously out of sync with the insect's traditional range. Cold winters and cool summers generally made the park an unsuitable home for the beetle. Tweedsmuir was a good six hours' drive north of beetle territory in the Chilcotin. Nevertheless, an infestation had popped up in 1995. Park officials responded by burning 1,500 acres, and then they burned another block. Four years later, officials flew over the park only to discover 250,000 acres of fading lodgepoles. After a reconnaissance mission through a forest as red as a sunset, Carroll felt both fear and defeat. "The beetle was already on a trajectory we could not influence," he suspected at the time. He didn't know it then, but he was also standing on just one of 22,000 beetle hot spots roiling across the province.

BRITISH COLUMBIA, a forest kingdom larger than France, occupies a particular place in the history of natural resource plunder in the West. For nearly a hundred years, logging has shaped the province's character and economy. Until 2004, wood accounted for 40 percent of all provincial exports and generated about $14 billion worth of revenue. To this day, the B.C. government calculates that its "pine dominated forest estate is a $240-billion asset"; as such, it says, "asset management is clearly required." But as H.R. MacMillan, the legendary forester, had predicted, an industry that began as a family-owned and locally based affair gradually fell into the hands of a few large corporations administered by lawyers "with a penthouse point of view" who had never had rain in their lunch-buckets. They turned the B.C. interior, a region dominated by spruce, pine and fir, into a stud-making empire designed to supply cheap wood to the U.S. housing market. Encompassing seven of the world's ten largest mills, the industry model focused on "cost minimization and volume maximization." So the beetle shook up a highly concentrated industry committed to producing large volumes of 2×4s. No company had a management plan or a lease agreement that made any allowance for beetle work. As one wag later put it, "Ten billion beetles can't be wrong—buy B.C. pine."

Most people living in British Columbia's lodgepole country still blame the government for not stopping what everyone calls the MPB, as though the beetle were a giant character from a Roald Dahl novel. People argue that if the New Democratic government of the time had allowed logging in Tweedsmuir, the plague could have been contained. Even intelligent and well-informed citizens claim that a socialist-leaning government prevented the necessary clear-cutting of a wilderness area to placate urban-based tree huggers. "The socialists elected to do nothing, and when the government changed hands the plague spread like wildfire" goes the myth.

But the truth is more uncomfortable and complicated. Only 8 percent of the province's mature pine grew in protected areas. The remainder still stand in timber leases controlled by industry. The government could have logged every one of the province's 972 beautiful parks and protected areas (that's 32 million acres) and still lost all of its lodgepoles to the MPB. By the time the outbreak peaked in 2007, it had reached such infernal proportions that the beetles annually occupied a landscape as large as B.C.'s entire park system (25 million acres). "We found a fingerprint in Tweedsmuir, but there were synchronized population explosions taking place right across the province up to 900 kilometers [550 miles] away," says Carroll.

Normally, a fading forest on Tweedsmuir's scale would have grabbed the attention of someone at Canada's Forest Insect and Disease Survey (FIDS). "I think FIDS would have noticed it by 1995 at least," says Carroll. But by then the federal agency no longer existed. Although trees cover a third of Canada, the world's largest exporter of forest products decided in 1996 that it no longer needed a national insect intelligence or forest health service. After sixty years of ably documenting the rise and fall of budworms, aspen ink spots, bark beetles, and the birch leaf skeletonizer, the federal government abandoned the agency to save money. As a consequence, it missed the beginnings of the continent's second-largest insect outbreak, a spectacular and unprecedented event that would cost the logging industry $50 billion.

The federal government had quietly passed the responsibility for monitoring forest health to individual provinces, and it took B.C.'s Forest Service several years to fill in the gaps. "There was no insect survey in '97 or '98, both critical years for increases in beetle population," recalls Carroll, who now teaches at the University of British Columbia. "In some provinces, still, no surveys are happening at all." It was as though the federal government "had poked out its eyes and blinded itself. We faced global change in the forest yet lacked the capacity to measure it," he says. At the time, Carroll's superiors told him not to talk about the information gaps or the demise of FIDS. "I was told to drop it."

FIDS had a novel and celebrated history. A group of scientists started the agency in 1936, after the European spruce sawfly managed to wipe out two thousand square miles of forest in Quebec's Gaspé Peninsula before anyone in government noticed. In short order, FIDS became the most important biological survey in Canada. Every year, FIDS technicians, loggers, and even rural police officers collected forest insects, performed aerial surveys, and made roadside or horseback observations on the extent of red-topped, green, or graying trees. The agency also documented rare and curious events in the forest. In 1962, for example, one report described a massive girdling of pine and spruce trees by a remarkable population explosion of hares around Prince George, B.C. The rabbits killed more than half the trees.

FIDS also kept track of the comings and goings of the mountain pine beetle in British Columbia's vast lodgepole forests. In 1943, the beetle killed 80 percent of the timber in Kootenay National Park over an area of 250 square miles. The 1951 report noted a "severe infestation in white pine stands along the Big End Highway at Downie Creek (Mile 42)." After the Chilcotin outbreak that Carroll witnessed as a child, a 1988 report observed that the pine beetle had harvested six times "the area burned by forest fires." In one of the agency's last reports, FIDS warned that the mountain pine beetle was "the most damaging insect of pine forest in B.C. and Alberta," because it had killed 230 million mature pine trees since 1972. In 1994, FIDS reported 9,340 infestations over more than 85,000 acres. The 1996 FIDS atlas of major forest pests, the last of its kind, even highlighted a startling increase in beetle activity in the "Pacific Maritime" near Tweedsmuir Provincial Park. But then the scientific light on the nation's largely publicly owned forests went out. "The government dropped the ball and was blind for two years," adds Rod Garbutt, a former FIDS technician. Like Carroll, he believes that good monitoring would not have changed the course of the beetle tsunami but might have rationally shaped and directed the multi-billion-dollar response.

After his Tweedsmuir visit, the B.C. park service asked Carroll what could be done. Once he'd consulted with federal colleagues at the Pacific Forest Centre, Carroll advised the park service that the best the experts could do was to predict the outbreak's size and determine what was driving it. He could use old FIDS data to figure out the particulars. The service "kind of said yes" to the plan, Carroll says, but then spent nearly a million dollars on a "fell and burn" program. They cut, peeled, and then burned individual infested pine trees at $100 a tree instead. "We had a chance to get going on understanding this thing, but the money didn't appear for another four years."

That's how the second-greatest insect infestation in the history of North America rolled over a landscape the size of California and brought down a forest that could fill most of Colorado. The epidemic changed entire watersheds and unsettled more than thirty rural and ninety-seven First Nations communities. Economic analysts say the beetle tsunami will influence the price and availability of pine 2×4s on the continent for the next fifty years.

In addition to confounding short-sighted politicians, the mountain pine beetle electrified corporate symposiums and highlighted the dismally unsustainable character of industrial logging. The outbreak underscored the appalling state of forest management in British Columbia. Just about everyone living in the province's interior—several hundred thousand people—lost trees, jobs, or income. Even civil servants prayed for the creation of a Ministry of Holy Crap to magically intervene and sort things out. In central B.C., aboriginal children started to color pines red instead of green. When wide-eyed Alaskans looked south and saw an outbreak ten times the size of their spruce storm, they gasped. "What happened in B.C., well, that was just unbelievable," says Ed Berg. And it was.

BRITISH COLUMBIA'S lodgepole forests are so ancient and ubiquitous that most aboriginal groups have bark beetle stories. The Tahltan, an Athabascan-speaking people who live smack dab in northern pine country, tell about the old days when people slept under rabbit-skin blankets. At that time Wormwood (beetle larvae) and Mosquito lived together. Every day Wormwood watched Mosquito fly about and come home happy and full of blood. Mosquito, however, didn't want to give away his secrets. So when Wormwood asked where he got his dinner, the sly Mosquito said he "sucked it out of a tree." *Dendroctonus ponderosae* still bores into the bark of mature trees looking for blood.

In his beetle bible, Stephen L. Wood had lots to say about *Dendroctonus ponderosae*, a beetle Andrew Delmar Hopkins, the father of North American entomology, called "the enemy" of pine forests. Wood devoted six full pages to "the most destructive species of Dendroctonus." He estimated that this scolytid, always led by female attackers, probably destroyed 1.5 billion board feet a year. Incredibly, the B.C. outbreak destroyed 43 billion board feet in a little over a decade. Wood wrote that the beetle attacked mostly weakened or large-diameter pines, including ponderosa

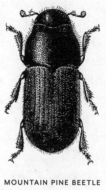

LODGEPOLE PINE

MOUNTAIN PINE BEETLE
(*Dendroctonus ponderosae*)

and lodgepole. But during the most recent outbreak, the pine beetle did everything the experts said it couldn't do: it flew over mountains, it invaded northern forests, it attacked spruce trees, and it wiped out pine plantations not much thicker in diameter than baseball bats.

The main object of the beetle's attention is the stately lodgepole, a tree that defines the Rocky Mountain west as definitively as do grizzlies. The lodgepole grows almost everywhere in the region and can be found from Alaska to Mexico. It tends to favor elevations of about six thousand feet, though it can thrive in any kind of soil. Walking up a mountain, you'll be embraced by Douglas-fir before you find lodgepole and then enter the splendor of Engelmann spruce. The lodgepole can grow in mixed-forest stands or as a single crop. It is sun loving and fast growing. At one time almost every Plains tribe, from the Blackfoot to the Cheyenne, regularly harvested lodgepole pines to make teepees. They also routinely fired pine forests to improve hunting or groom trails.

The lodgepole, once considered a weed, didn't become a global commodity until thirty years ago. After logging companies wiped out most of the valuable and mature spruce and Douglas-fir trees throughout the West, they cast about for smaller stock, the same way the fishing industry started to net smaller fish. In the 1980s, the U.S. Department of Agriculture did a study on the tree's anatomical, chemical, and mechanical properties, with the specific goal of "improving utilization of lodgepole pine forests of the 21st century in North America." The study found that one in four trees west of the Mississippi were lodgepole. While *Pinus concorta* occupied 12 million acres of U.S. forests, the lodgepole occupied nearly a quarter of western Canada's timberlands. But there was one drawback, the study said: "A significant portion of the forests were dead." In 1979 alone, the infamous MPB Logging Company had taken out well over a million acres.

No conifer has a more intimate and intense relationship with a scolytid than the lodgepole does. By targeting mature pines between 80 and 120 years of age, the beetle helps the pine forest prepare for renewal by fire. As every aboriginal person once knew, and the white ecologist Fred Clements first learned in 1910, only a good fire can open the closed-fisted cones of older lodgepoles to release their prolific seeds. (Young trees are not so stubborn.) Fire not only prepares a good seedbed, it also opens up the land to sunlight and removes all seedling competitors on the ground. Fire keeps a check on beetle populations, too, which explains why many lodgepole forests look so similar in age.

In 1987, the entomologist Ken Raffa proposed in *American Naturalist* that the MPB managed lodgepoles like some sort of radical shepherd. In the careful way scientists word things, Raffa suggested that the beetle slowly raised and then destroyed a distinct pine monoculture with much greater finesse than did most logging companies. After looking at several kinds of bark beetles, Raffa reasoned that the MPB employed a unique set of behaviors and skills. Unlike, say, the fir engraver, the MPB seemed to wait patiently for healthy trees to reach eighty to one hundred years of age. At that point, more than two-thirds of the trees can no longer defend the castle with resin resistance.

The creature's mass attacks were also highly strategic. Compared to the fir engraver, the MPB carefully limited the density of attackers to seventy per square yard—just what was required to kill and condition the lodgepole for optimum beetle reproduction. It also minimized overcrowding in the nursery to make maximum use of the phloem. Severe pine beetle outbreaks selectively created fuel for a seed-opening fire, which in turn prevented the local extinction of the lodgepole. Raffa concluded that the lodgepoles actually let down their guard in old age to increase "the likelihood of the beetle epidemics and subsequent fires, thereby favoring reestablishment of lodgepole pine."

After reading Raffa's paper, many researchers thought the philosophical entomologist was a lunatic. People who have lived in the forest all their lives, however, know differently. Early in his career, Raffa worked in a logging community in eastern Oregon. One day a logger asked him what he was doing. "I'm here to protect the trees from beetles," said the academic. The logger laughed and said that was bullshit: "The trees and the beetles have been in cahoots for millions of years." Raffa remains moved by this man's eloquence.

Raffa had really spelled out a dynamic that nobody in authority wanted to pay attention to. At least, that's what Allan Carroll found in 2001 as pine beetles exploded over two and a half million acres of forest. Rather than wait for special funding ($40 million for research didn't arrive until 2003), Carroll mined old FIDS data to discover the most critical driver of MPB outbreaks throughout the North American West: the whole beetle, lodgepole, and fire dynamic had been thrown off kilter. "It's a perfectly tuned, complex system. We gave it a bit of a jab and buggered it up," says Carroll.

In the early 1900s, only 17 percent of the lodgepoles in British

Columbia were old enough to be attacked by the MPB. (In an unmanaged forest, wildfire typically keeps the percentage of beetle-mature pine at around 25 percent, so there is lots of resilience.) But after foresters started to suppress fire to protect the province's timber wealth, the volume of mature lodgepole increased, to 53 percent, or 20 million acres, by 1990. The forest looked more and more like a national or household economy, bearing enormous debt and incalculable risk. Between 1910 and 1990, four beetle outbreaks erupted to manage this extreme overdraft. The outbreaks increased in intensity and area over time. British Columbia's forests were an accident waiting to happen.

Fire suppression, a genuine western fetish, began at the turn of the twentieth century, when wood became "the mainstay of the Provincial Treasury." Because a burning tree represented a lost dollar, fire protection amounted to posting armed guards at a bank. During the 1930s, the Canadian government even forbade traditional burning by aboriginal nations. By the late 1950s, the Forest Service boasted a heavy-equipment arsenal of fire-line plows, bulldozers, and fire pumps and its own air force of helicopters and water bombers. Costs soared as the anti-fire machine jumped on 90 percent of fires from four acres in size up and snuffed them out. The multi-million-dollar annual effort created a more uniform, dense, and expansive patch of aging lodgepoles.

The scenario reminded Carroll of a lesson he had learned in Canada's eastern forests. While battling spruce bud moth in single-age spruce plantations, the ecologist realized that human engineering that reduced a forest's diversity, either in species or in age groups, created a landscape vulnerable to catastrophic change. Public policy in British Columbia had turned a lodgepole forest once diverse in age into a monoculture of aging baby boomers. Nearly thirty years earlier, foresters had identified the same pathology in East Texas, when the southern pine beetle went on a roar. "What we have here is not an epidemic of southern pine beetles but an epidemic of southern pine," they said.

At first, many professionals didn't believe that fire suppression removed resilience from the forest. But over time, they had to face the hard facts. (In B.C., a 2003 firestorm of 2,500 blazes that consumed 334 homes, forced the evacuation of 45,000 people, and burned through 540,000 acres graphically illustrated the downside of successful fire suppression in an ecosystem that depends on fire.) "Economically, it may have made sense to remove fire from the landscape, but as a consequence we grew this massive inventory of mature pine," explains Carroll. "We created a fairly substantial pine beetle smorgasbord." By religiously stomping out fires in Montana, Oregon, California, Colorado, and Wyoming, foresters and park wardens there also prepared the ground for both explosive insect insurgencies and extreme wildfires. In Grand County, Colorado, for example, foresters introduced strict fire control after railway barons cut down nearly 70 percent of the lodgepoles in and around the 1880s. More than one hundred years later, a buffet of mature pine awaited the native pine beetle. Foresters in Colorado prayed for a cold winter, but the weather did not oblige, and beetles killed more than 80 percent of the mature pine. With subtle variations, the story played itself out on various scales all throughout the West.

As the scale of B.C.'s outbreak grew from 2 million to 10 million acres by 2003, Carroll suspected that something else was happening in the forest. Dedicated firefighting may have set the table, but it looked as if climate change had reconfigured the table's shape and size, so that it now extended into northern and alpine forests. After receiving more reports of beetle activity throughout northern B.C. and even in northern Alberta, across the mountains, Carroll again mined FIDS data and examined provincial climate records for clues.

He soon found exactly what Ed Berg discovered in Alaska: a subtle systematic warming that freed the beetle from all cold restraint. An insect that once had had to work for a living now lived large, like some hedonistic teenager with an unearned inheritance. Over the last hundred years, average minimum temperatures in central B.C. have increased by nearly six degrees on the Fahrenheit scale. As hotter summers stressed out the forest, warmer winters put more young beetles on the landscape. Normally, only 20 percent of beetle larvae survive a classic Canadian winter. But the last -40-degree winter day—the minimum temperature needed to assure beetle kill—had occurred in the 1980s. Now 80 percent of the beetles came out flying. At the same time, climate change had shifted suitable habitat for the beetle by more than seven degrees latitude north over the past thirty years. In the absence of a cold snap, Carroll predicted at the time that "the current outbreak may not entirely collapse as in the past" but would continue as it moved into northern populations of pine. More warming, he said, would extend the beetle's empire "northward, eastward and toward higher elevations." He also suspected that the beetles would fly over the Rockies into a lodgepole plateau in north-central Alberta and thereby enter virgin territory: the jack pine in the boreal forest. Unlike the majority of economic predictions made in 2004, every one of Carroll's biological forecasts came true.

At first, no one in either the Canadian or the British Columbian government wanted to hear about climate change. One assistant deputy minister at Natural Resources Canada reprimanded Carroll at a public meeting, charging that anyone tying the beetle outbreak to climate change "was doing so without proper scientific basis." Carroll repeatedly attended beetle congresses and summits where he was the only participant who openly talked about "the climate-change trigger."

The forest industry was scrambling. In many of the province's eleven forest districts, the beetle was out-logging sawmills twenty-three to one. In 2001, in a response, B.C.'s chief forester, Larry Pedersen, upped the allowable cut, in some cases doubling or tripling the number of trees companies could pull off the landscape. "This is all about salvaging the dead wood," said Pedersen at the time. "We are expecting these timber supply areas to be over-run. Vast areas of the forested landscape are dead, and there is a one-time opportunity to capture the economic value." Simultaneously, the B.C. government lowered its stumpage fees, or tree royalties, to a give-it-away price of 25 cents per cubic meter. Industry spent millions expanding the capacity of their "super mills" to handle the surplus beetle kill. By hauling green beetle kill all over the province, loggers created mini-epidemics along the highways. Heavy logging traffic also destroyed nearly a billion dollars' worth of public roads.

Around the same time, the province commissioned a quick review on the costs and benefits of clear-cutting beetle kill. Not surprisingly, the review found that huge clear-cuts accompanied by new logging roads reduced a forest's complexity and diversity. Letting dead trees stand, by contrast, allowed "the natural processes of death, decay, regeneration and maturity to continue and provides the habitats for species dependent on those processes." Bumping up harvest rates, said the review, would only increase "the risk of negative impacts to the diversity of wildlife and wildlife habitat." Politicians ignored the review's findings.

The beetle epidemic caught B.C.'s Forest Service in a skeletal state. Since 1995, government budget cuts had reduced forest service staff by 40 percent. In particular, the division responsible for counting trees had suffered an 85 percent clear-cut. As a consequence, the government's ability to "adequately and consistently measure and quantify the level of beetle infestation across the province" was in doubt for much of the outbreak. Organized groups of forest professionals told the government in 2004 that "without a solid foundation of information it will become increasingly difficult for our members to make sound, science based, forest resource management decisions." The upbeat 2001 *Report of the Mountain Pine Beetle Task Force*, however, viewed the beetle "emergency" as opportunity to create as many tree-felling jobs as possible and "turn a potentially negative situation into a positive experience." Letting nature take its course, the report said, was "not a valid option."

To that end, the province appointed a beetle boss, Bob Clark, to oversee action plans. It also created a beetle coordination committee. After its fell and burn programs didn't solve the problems, the province placed its hopes on "snip 'n' skid," in which loggers cut down small patches of infested trees and then dragged the wood to logging roads using a towing machine called a skidder. Over a three-year period, the government performed 15,000 snip 'n' skids. "It was probably a defense unparalleled in the history of man," Bob Clark later told *The Walrus*. When the snipping failed, government gave industry permission to create monster clearcuts in advance of the beetle hordes. But the beetles flew over or around them. Clark rationalized the battle by telling a local entomologist that "just because half the people die doesn't mean you close all the hospitals. You have to make an effort."

Sometimes the effort included out-and-out poisoning. Industry called it "hack and squirt." Between 1995 and 2004, beetle battlers applied nearly 9 tons of arsenic or monosodium methanearsonate (MSMA) to some 500,000 trees. Pest managers would hack a wound in the lodgepole's bark and then squirt in the arsenic. The phloem drew up the poison. A fell burn operation cost $75 a tree, but arsenic poisoning cost only $30 per tree. "You could do four times more trees than fell and burn in a day," explained one entomologist.

When a citizen in Smithers, B.C., raised questions about the risks of arsenic treatments near private property in 2001, the province's Forest Practices Board reluctantly launched an investigation. Josette Wier, a French-born pediatrician, pointed out that arsenic was a carcinogen, a DNA changer, and that forestry workers using the poison often had high levels of the stuff in their urine. Three years later the board, an Orwellian body, acknowledged that there was some debate about the safety of poisoning beetles with arsenic. It noted that the provincial Forest Practices Code strictly forbids poisoning water, land, or animals, but it ruled that "the prohibition against damage to the environment does not apply to the use of pesticides."

The province didn't suspend its arsenic treatments until a 2006 Environment Canada study showed that arsenic was killing Mother Nature's best beetle controllers: woodpeckers. (One B.C. government study awkwardly referred to the 15 species of woodpeckers in question as "bark-foraging wildlife tree users.") Woodpeckers can slow down an outbreak or speed up its demise. The stomach of one three-toed woodpecker can hold 915 beetles or beetle larvae, and the bird has to fill up several times a day. During a beetle outbreak, woodpecker densities can increase up to 85-fold in the forest. It's not uncommon to find as many as 12 woodpeckers working the same lodgepole. By flaking and removing bark, the woodpeckers expose beetle larvae to winter cold and significantly increase beetle mortality. The federal study found that the practice of hack and squirt resulted in "significant accumulation and transfer of organic arsenic within the food chain." Many woodpeckers foraging around poisoned trees developed levels of arsenic high enough to cause weight loss and death. The study also found that industry claims about MSMA's ability to kill 90 percent of beetle larvae had been grossly overstated. In reality, the arsenic killed only about 60 percent of the beetle population—not enough to make any difference.

The outbreak that flummoxed all the usual experts also broke every beetle record imaginable. "The two things that surprised me and even embarrassed me to some extent were the size of the population and how fast it built," says Staffan Lindgren, a beetle expert at the University of Northern British Columbia in Prince George. Historically, about six hundred beetles would attack a tree. But during what Lindgren dubbed the "hyperepidemic," as many as six thousand insects overwhelmed individual trees. "In some cases, it was a suicide mission. They had no room to breed." The scolytid had even changed its flight times. It normally hunted for a new home in July and August, but now the beetles emerged early in June and popped out of trees all summer long. Traditionally, researchers gauged the spread of an outbreak by mapping changes in the color of the tree canopy from the air. But the beetle confounded that practice, too. Swarmed trees normally remain green for a year before they started to fade, Lindgren explains, "but the trees now faded in July. They died immediately. By fall they were red. Our system for figuring out what was happening went out the window." One of Lindgren's PhD students also discovered that virgin lodgepoles in the north didn't know how to repel the attackers. "They didn't respond with the right chemicals."

To make matters worse, just as the outbreak was taking off, British Columbia's Forestry Ministry had deregulated and transferred responsibility for monitoring large chunks of land to multinational lumber companies. The new policy required that competing timber barons work together to survey their country-sized domains. "In a number of cases it didn't work," says Lindgren. "The monitoring was dropped on industry's lap while they were in panic mode. There were a lot of coincidences that didn't help."

WHEN THE beetles started to invade Prince George, a logging community of 80,000 people, Mark Fercho, the city's environment manager, tried to play by the rules. He thinned stands of trees in parks and put up pouches filled with verbenone, the chemical perfume that beetles use to tell other beetles a tree is full. Search-and-destroy crews patrolled the streets looking for red-topped trees. Fercho avoided pesticides, because he knew they'd never improved the nervous systems of dogs or fish.

But as the flights intensified in 2003, nothing seemed to work. Thinned stands got clobbered, and beetles boldly drilled underneath the verbenone pouches. While beetle and fire crews checked for pitch tubes at the base of trees, incoming beetles landed in the canopy and bored into the top third of the tree. The beetles, which were supposed to prefer twelve-inch-thick pines, now riddled poles less than four inches in diameter. At that point, Fercho threw out the beetle rulebook. "When you get punched in the face first, you have to deal with it."

He refocused the city's removal program on mature trees and fire hazard control. He sold the affected wood to local mills and invested the money in the city's salvage program. Many parks ended up looking like "plucked chickens," Fercho says. The Pinewood golf course and Pinewood school both lost their namesakes. "When we started to take out green attacked trees, the opposition was tremendous. I got beat up so many times. But once the trees turned red, we couldn't get rid of them fast enough. That was really striking."

By 2004, beetles were arriving all summer long and into the fall. "One huge flight came in on Halloween, at the end of October," Fercho remembers. "Nobody had ever seen that before." He often arrived at a park or a golf course just after the trees had been hit. "You could hear the beetles chewing and see the sawdust falling down. You felt for the trees. It was creepy and scary." Every day, local radio stations announced where the beetle and fire fuel crews would be working. Over a three-year period, the city filled more than six hundred logging trucks with sixty thousand dead or dying pine and spruce trees. At one city council meeting, Fercho explained the philosophy behind the most successful case of urban beetle response in North America: "We are a cork on the ocean. We can't avoid the wave. But we can determine how gracefully we want to bob up and down."

IN 2005, the government-created B.C. Market Outreach Network put up a sign on Highway 20 in Chilcotin country at the Hanceville Lookout. By then, the forest looked as if someone had thrown a bucket of red paint on it. The sign declared that the MPB had infested 7 million hectares—17 million acres—of interior forest, an area the size of New Brunswick, and would change the lives of 25,000 logging families. A decade of mild winters had reduced mortality rates among beetles from 80 to 10 percent. Government could trim, cut, and survey the forest, the sign explained, but "they can't stop a massive epidemic like this one." In the end, "while managerial techniques can help, nature will always play a critical role in balancing the forest ecosystem."

Nature also taught Robert Hodgkinson a thing or two. He's the forest entomologist for B.C.'s Northern Interior Forest Region, an area the size of Norway. Given the wealth of mature pine and the disappearance of cold winters in the region, Hodgkinson recalls, "Our hazard rating looked like a chart for an old guy at risk of a heart attack." Although crews harvested green timber and applied arsenic, neither tactic made much of an impression. The beetles came on in an endless wave that moved as fast as forty miles a year. Hodgkinson found beetles reproducing in trees as small as four inches in diameter. "We won a few battles but lost the war. Every year we had to back off and retreat on the landscape. We were like the Germans in Russia."

One of the rules in the old beetle book was that the MPB wouldn't dine on spruce, let alone reproduce in a different species. But in 2006 Hodgkinson found twelve spruce trees attacked by pine beetles just outside of Prince George. Two of the trees were dead. "There was living larvae in those spruce trees. We'd never seen that before." The interlopers were even producing more offspring per pair than were breeding adults in nearby pines. Given time and a warming climate, it was clear, the MPB could switch hosts and develop into a spruce-seeking species.

Hodgkinson battled the beetle in his own Prince George backyard, where three majestic lodgepoles then stood. He attached verbenone pouches on two trees, but after ten days the trees sported telltale pitch tubes. "There were so many beetles flying around that verbenone didn't have much effect at all." Next he put on protective gear and sprayed the trees with Dursban, an organophosphate that blasts the nervous system. He even let the bug killer run down to the base of the trees in gobs. Soon a "ring of death" appeared at the base of the trees, but "the beetles still kept coming," says Hodgkinson. "They climbed over dead beetles. I could hear them chewing in the trees. I killed tens of thousands of beetles with insecticide, but they still kept on coming." Like most residents of central B.C., Hodgkinson lost his pines. About the same time, journalist Ben Parfitt wrote that stopping the beetles "would be as easy as sucking up an incoming ocean wave with a straw."

In 2004, the government had asked Allan Carroll to check out more dead trees outside Chetwynd in northern British Columbia. Chetwynd is a small town just an hour's drive from the Alberta border, about two hours from jack pine country and the boreal forest. Over a nine-hour helicopter trip, Carroll repeatedly landed to examine dead or dying trees. He'd pull out his bowie knife and scrape away the bark. He invariably found larvae and J-shaped galleries, even though there had never been any record of the MPB that far north. Carroll figured that the insects had come from the center of the outbreak, a good eight-hour drive south, and had flown over the Rockies. The B.C. government declared another "Emergency Bark Beetle Management Zone" as aboriginal crews frantically felled and burned trees.

As the beetle empire moved north, Peter Jackson, a meteorologist at the University of Northern British Columbia, started to see odd blips on the equipment at the newly installed weather radar station on Mount Baldy twelve miles south of Prince George. The blips "looked like a light drizzle or light rain," he remembers. The low-interval echoes clustered at low altitudes and "lasted for hours and hours." He thought they might be caused by insects but wasn't sure. The clouds surged between ten in the morning and ten in the evening, a period in which temperatures climbed above 68 degrees Fahrenheit. Beetle experts had long suspected that *Dendroctonus* could catch updrafts and travel great distances, but no one had caught them in the act. "There was little known about long-range dispersal," recalls Jackson. "Everyone suspected they were clumsy weak flyers and could only remain airborne for a few hours."

In 2005, intrigued by the radar findings, Jackson rented a Cessna and attached one butterfly-like net under the wing. During the months of July and August he had the pilot fly a transect south of the city at different altitudes. At 15-minute intervals, Jackson pulled in the nylon net and put out a new one. He netted as many as 40 beetles per transect one afternoon. "We got some big numbers." That meant about 850,000 beetles per hour were flying above the canopy across a 330-foot line at right angles to the wind. By catching winds at 12 miles an hour, the beetles could motor more than 60 miles if they remained airborne for just 5 hours. Jackson calculated that there were enough beetles in the air every 330 feet to kill more than 1,400 trees an hour. (It takes about 500 beetles to kill one lodgepole.) To his astonishment, he found beetles flying as high as 2,600 feet above the forest canopy.

AFTER THE MPB crossed the Rockies, beetle hysteria gripped the government of Alberta. Although the bug had entered southern lodgepole forests in the past, the prairies had served as a natural barrier to any continental expansion. Thanks to a century of fire suppression, nearly $28 billion worth of mature lodgepole occupied the southern slopes of the Rockies; there was nowhere for the beetle to go after that. Northern Alberta, however, was another story. The lodgepoles of Peace River country eventually melted into the boreal forest, where jack pine dominates. The jack pine stretches all the way to Newfoundland and even connects to loblolly pine forests in the eastern United States.

For Canadians, the jack pine symbolizes tenacity and perseverance. After Group of Seven artist Tom Thomson painted a lonely jack pine by a northern lake in 1917, he died in a mysterious canoeing accident. But his pine portraits *The Jack Pine* and *The West Wind* became two of the country's most famous icons. Each portrays a solitary jack pine seemingly growing on rock by a startlingly blue lake. For Canada to lose its jack pine would be like Lebanon losing its cedars.

Alberta reacted to the threat by burning taxpayer funds, nearly half a billion to date. To remove just 5,000 infested lodgepoles near Jasper National Park, the province employed an army of 300 aboriginal workers and 14 helicopters. Staff at Parks Canada took a different approach, with a $7-million prescribed burning program that targeted prime MPB forest habitat around the town site of Banff, where fire has been religiously suppressed since 1875. Between 2002 and 2006, Parks Canada burned nearly 50,000 hectares of lodgepole pine in an attempt to give the MPB less dining room. According to Bill Fisher, a Parks Canada administrator, the burns not only deprived the beetles of a corridor but also restored bird abundance, bear density, and species richness in the rapidly greening burned areas. Adds Fisher, "The problem is bigger than a single province and a single ecosystem."

Allan Carroll spent much of his time accompanying Canada's minister of natural resources, Gary Lunn, on tours of the attacked forests in Alberta and British Columbia. Carroll recalls one meeting in Alberta at which senior politicians and civil servants spent more time discussing the glories of big game hunting than they did the beetle emergency. Whenever his phone stopped ringing, Carroll sat at his desk, contemplating a model of the pine beetle that was about the size of a large dog. Just about everything about the insect in its recent incarnation defied common sense. It was normally a bit player in the woods, a poor competitor and a terrible flyer. Most bark beetles wisely avoided risk. But not the MPB. It attacked large, living, resin-rich trees. Unlike most of the

Dendroctonus clan, it flew late in the year, often in July and August, putting its larvae at risk of early frost. All of its beetle competitors looked for "a sure deal," but Carroll concluded the MPB was a "gambler looking for an opportunity." Here was an extreme insect with a winner-take-all mentality.

In May 2006, beetle mania moved outside of rural Canada to briefly occupy an urban stage at the Calgary Hyatt Regency. There, a bevy of politicians, deputy ministers, scientists, and forest managers from Alberta and British Columbia gathered for an urgent and unprecedented "summit" on the opportunist in the forest, which one politician disdainfully referred to as an insect "no bigger than a mouse turd." Inside a cavernous hall equipped with several large screens, one PowerPoint presenter after another showed alarming maps of the beetle's impeding invasion of the boreal forest. Dave Coutts, Alberta's minister of sustainable development, declared that it was critical that people understand the urgency of the situation and join together to "combat this pesky little bug." According to Coutts, only Alberta's watchful foresters now stood in the way of continental disaster.

To convince Albertans and other westerners of the outbreak's magnitude, the Calgary summit instructively began with an aerial tour of the war zone in mountain pine beetle country. The six-hour round trip between Calgary and Williams Lake, B.C., flew over six forest districts under various stages of attack. As B.C. MLA John Rustad later explained to a host of beetle tourists during a salmon lunch in Williams Lake, "It's not every day you get to see a natural disaster in slow motion." On board a twin prop Dornier 328, Allan Carroll, looking exhausted and drawn, described clearly how swarms of beetles salt-and-peppered the forest and then filled in the landscape. "First, the trees turn red. Then they shed their needles and turn to gray."

The first signs of colorful die-off appeared just north of Kamloops, where beetles had infested 20,000 acres. Carroll pointed out that a beetle attack turns ponderosa pines orange but colors a lodgepole bright red. From the air, the infestation at first looked like a hit-and-miss affair. But that sense of haphazardness soon gave way to well-demarcated attack zones illuminated by garishly red hillsides in the 100 Mile House Forest District, where beetles had already consumed 220,000 acres. Farther north, in the Chilcotin Forest District, red and gray vistas had replaced green vistas altogether. Huge cut blocks appeared in the ghostly gray forest where crews had frantically logged "beetle wood" before it rotted.

Around Williams Lake, the whole country dissolved into a ghostly red hue. Pointing to an endless sea of crimson and ocher pines, Carroll talked about "the heart of the big red blob." By the time the plane flew over the Cariboo and Quesnel forest districts, the devastation brought to mind an unforgiving biblical plague. "The beetles have eaten just about everything," said Carroll. In Quesnel, that meant the beetles had marched through two-thirds of 2.3 million acres of commercial pine. An Albertan on board the flight asked if anything could be done. Carroll replied that you couldn't do much about an outbreak this size: "A bucket of water will put out a campfire but not a burning house."

Back at the summit, a variety of beetle veterans summed up the state of things. Sounding like chastened stockbrokers, many emphasized the dangers of holding a forest portfolio heavy in mature pine; others praised diversity. Doug Routledge, vice president of B.C.'s Council of Forest Industries, told everyone that "we did too little too late." Rich Coleman, then B.C.'s minister of forests, admitted that the province had created the proverbial perfect storm: "I can't stop the infestation in B.C. It's too big. Too massive. It can move at a rate that is unprecedented. We have trillions of beetles, and they're tough little nuts to deal with." Keith Dufresne, manager of the Cariboo-Chilcotin Beetle Action Coalition, suggested, "If we had had a Ministry of Holy Crap, we might have done a better job." Dufresne added that Canadians should prepare for the unthinkable in the boreal forest. "It's not something you've ever seen before. It's like a tsunami that takes twenty-five years instead of two seconds."

A few speakers at the summit noted that the forest wasn't really dead. Rather, the beetle was renewing an aging pine forest in an uncomfortable time frame for logging companies and tourists. Dave Neads, a conservation consultant and member of B.C.'s Mountain Pine Beetle Task Force, warned about the vulnerability of forest monocultures and the human communities dependent on them. "If you have diversity in your forest," Neads said, "communities and economies can withstand stress." He pointed out that rural communities that diversify their economies will be better prepared for the biological and political grief to come. "As global climate change proceeds, this MPB might be the first of many invasions. We have entered the rapids." At the end of the two-day summit, Coutts, the sustainable development minister for Alberta, solemnly asked the crowd to get down on their knees and pray for divine intervention. "Pray for cold weather. Pray tonight . . . This epidemic could come our way if our prayers don't work."

Despite all the praying and an expenditure of $200 million on fell and burn crews, a huge flight of beetles landed throughout Alberta's Peace River country in 2006. After riding thermals for 190 miles from British Columbia, the insects fell like rain out of the sky. Farmers heard them ping on the roofs of metal barns. Hungry and fatigued, the beetles quickly hit more than 300,000 lodgepole trees, exactly where the Rocky Mountain West meets the great boreal forest. On each tree, they produced enough offspring to attack another ten. Hardly a national newspaper carried the story. It was, after all, just a story about trees. Allan Carroll, however, recognized it as an "unnerving and disturbing development." The beetles had crossed a new threshold and tipped another point. "We have never seen beetles there before."

AFTER STUDYING group attacks on pine trees over a thirty-seven-year period in B.C., Dr. Javier Gamarra at Aberystwyth University in Wales concluded that mountain pine beetles behave like guests arriving at a celebration. "By observing people as they arrive at a party," Gamarra wrote, "you find that the early arrivals will group together." Over the evening, the difference in size between the larger and smaller groups grows. A shy, lonely soul might end up in the corner with a drink as a companion while the popular guy joins the big groups and parties on. Cold winters put "lonely souls" on the landscape, but by increasing larval survival rates, mild winters make for wild, "popular guy" summer beetle parties in drought-stressed forest. "The bad news is that climate change will produce increasingly milder winters reducing larval mortality and raising the risk of damage to many forests and crops," according to Gamarra: party time for beetles.

Researchers have also studied the characteristics of MPB pioneers. The complexity of the insect's decision making startled them. Pioneer beetles tend to die at higher rates and reproduce poorly compared to the rest of the swarm. Yet the pioneers, who arrive first and recruit helpers for a group attack, are neither overly fit nor in poor shape. They tend to be ordinary females of normal body weight with normal levels of energy reserves. Pioneers can't afford to sit back and wait for things to happen. They tend to attack smaller trees early in the season, so that if they choose wrong, they have enough energy to scale another castle. No one has determined yet why pioneers select the pines they do.

The famous western painter Charles Marion Russell never truly liked "pioneers" or being called one. Pioneers basically killed the West, Russell often explained. "In my book a pioneer is a man who comes to virgin country, traps off all the fur, kills off all the wild meat, cuts down all the trees, grazes off all the grass, plows the roots up and strings ten million miles of wire. A pioneer destroys things and calls it civilization." A pioneer beetle performs in a similar way.

In 2010, Allan Carroll attended the annual Western Forest Insect Work Conference in Flagstaff, Arizona. About a hundred foresters and entomologists gathered in the Aspen Room at the Little America Hotel to hear him talk about the latest in unprecedented pine beetle developments. The ecologist matter-of-factly told his colleagues that beetles invading virgin northern forests in Alberta reproduce more successfully than those working in southern B.C. forests accustomed to scolytids. With no memory of beetle attacks, the northern lodgepoles failed to express resins or mount a dedicated defense. Like the Lakota and the Cree when European smallpox invaded their communities, the beetle simply overwhelmed the trees. These northern pine nurseries produced 50 percent more offspring than southern trees, Carroll noted. Tellingly, they also made more determined pioneers.

FOUR

The War against the Insect Enemy

. . . . .

"The eradication of any native species of forest insect is impossible. They are as much a part of the forest as the trees themselves. They have been there as long as the trees have, and will continue to be so until the end of time."

HECTOR RICHMOND, *Forever Green*

BRITISH COLUMBIA'S all-out war against the bark beetle, and the similar campaigns that erupted madly throughout the U.S. Rocky Mountain states, have their roots in Europe. Nearly three hundred years ago, Europeans invented the search-and-destroy tradition with the highest of scientific ambitions: saving more wood for humans. The continent's anti-scolytid crusades, which consisted of felling all infested trees, rapidly transformed the region's forestry schools into a training ground for beetle slayers and the continent's forests into battlefields. Europeans also exported their pest-centric thinking across the Atlantic, where New World foresters took up the creed with gusto. On this side of the ocean, the beetle warriors resorted to even more extreme measures, experimenting with electrocution, drowning, scores of bug-killing chemicals, and even an early version of napalm. Really creative researchers tried explosive charges and costly beetle-attracting perfumes. The beetle expert and acclaimed chemical ecologist John Borden acknowledges that the campaign to eradicate the bark beetle remains one campy history. It illustrates our "inadequate ingenuity and desperation," he says. And much more.

THE FIRST full-scale, no-holds-barred campaign against the bark beetle broke out in German forests. Its chief general was Johann Friedrich Gmelin, a doctor, botanist, chemist, naturalist, and remarkable polymath. Gmelin led the charge with the publication of *Abhandlung über die Wurmtroknis* ("Treatise on the Worm Dryness") in 1787. "No pests have ever done so much harm to the woodlands as has the bark beetle," wrote Gmelin. "It is therefore worthwhile to inspect him more closely." Gmelin's startling inspection filled more than five hundred pages.

Gmelin's interest in bark beetles was no accident. During the eighteenth century, the emerging German nation gave birth to the field of scientific forestry, which in turn supported elaborate studies on how to kill bark beetles. Although the country once boasted Europe's darkest and densest forests, a peasant population boom in the eleventh through the thirteenth centuries cleared, thinned, or grazed most of the broad-leafed trees. Historians call it the Age of Great Clearings. Pioneering Slavs and Germans, aided by agrarian Catholic clerics, dispatched oak and birch trees the way the buffalo hunters killed American bison. Around 900 AD, most of central Europe lay under a forest, but by 1350 peasants had chopped down nearly half of the forest. The historian Michael Williams calls the leveling of European forests one of the world's "great deforestation episodes." Only the Black Death saved the few remaining old-growth stands from the pioneering depredations of *Dendroctonus homo economicus.*

After the plague emptied nearly a quarter of Europe's villages, spruce and pine trees recolonized much of the land. But by 1650, another burst of human activity placed Europe's surviving forests in peril. In some regions, the aggressive wood appetite of shipbuilders, iron smelters, and city builders created severe shortages or outright tree famines.

The solution was to industrialize tree making. Men of commerce and science decided that woodlands could no longer be left in the trust of peasants who depended on community forests for fuel, fertilizer, and food. Europe's endangered trees, they argued, needed a professional class of guardians in the name of "dear posterity." The founders of modern forestry, such as Hans von Carlowitz, Georg Hartig, and Heinrich Cotta, proposed an innovative forest economy based on exact measurements, single-species high-yielding tree plantations, and the new gospel of efficiency. "Order was imposed on disorderly nature, geometric perfection replaced the ragged edges of natural growth and sustained yield became the forester's guiding principle," notes Williams in his monumental history *Deforesting the Earth*.

The scientific foresters saw themselves as enlightened and sustainable tree farmers. In their hands, trees, once considered "a gift of nature," became another crop. This new educated breed introduced the practice of staggered clear-cuts as well as the planting of fast-growing species such as Norwegian spruce and Scots pine. As rural communities lost their hunting and gathering rights in the woods, tree professionals gained access to predictable, measurable cords of wood from tidy plantations. The pioneers, in this case professional foresters, changed the face of Europe: a land once dominated by beeches, chestnuts, ashes, and oaks (and their assorted insect pruners) became a plantation of spruce and/or pine trees with acidic soils. Not surprisingly, these manmade forests attracted a horde of powerful critics: bark beetles.

By the late eighteenth century, the spruce beetle, *Ips typographus*, had already highlighted imbalances in the continent's impoverished forests. When Europe's wild places contained a diverse assortment of hardwoods and conifers, bark beetles hadn't made much of an impression. German peasants called some bark beetles *Wald gartners*, or forest gardeners. But after the medieval clearings, people couldn't help but notice the mass spruce die-offs in their wood-starved communities as the beetles swarmed mature crops of the trees.

Peasants called the spruce beetle "the flying worm," the black worm, or simply "the curse." By 1705, German peasants in the Harz region even had a hymn in their songbook to protect their forests from "windstorms and harmful worms." The fuel and charcoal demands of mining operations had devastated hardwoods, along with most wildlife in the region, so all that was left were uniform plantations of spruce. When winds blew down these shallow-rooted conifers, the beetle behaved as it would centuries later in Alaska: it turned undefended castles into fertile nurseries and went wild.

Bark beetle epidemics and wood scarcity grew so grim by 1782 that the Royal Society of Sciences at Göttingen offered an award for the best proposals on how to control the beetle. Gmelin, an eighteenth-century geek and Göttingen resident, answered the call with his *Treatise on the Worm Dryness*. An old-fashioned observer and naturalist, Gmelin didn't leave a twig unturned. He first marveled at the beetle's hardiness. The fully developed beetle "often lies dormant underneath the bark, encased in ice, over the entire winter as if dead. But the warmth of the breath, or a hand, or from an oven, and sunshine, makes him right again." The animal's pack behavior shocked the scientist: "These swarms of beetles are so burdensome for the people...they are soon covered with them and get a nasty itch from them."

Gmelin duly noted that beetles sometimes attacked only individual trees, causing "disorder in the forest household," but usually raged like a contagious disease causing "devastation" throughout the forest. The beetle attacked not only wind-felled trees but also healthy, mature spruce, in which the insect caused a peculiar sickness (the work of fungi). Gmelin alternately described the sickness as worm dryness, tree dryness, or spruce cancer: "This type of dryness is different than any other kind: just as a deadly pneumonia is different than a slow, creeping, pulmonary consumption."

In the Harz region, the beetle devastated such large tracts of forests that authorities mobilized soldiers to chop down the dead trees. The local ironworks got orders to use only "worm-dried wood." In 1782 alone, the curse killed 259,106 trees. Gmelin astutely noted that climate appeared to drive the outbreaks: "Sustained hot and dry weather...leads to the rise of the curse."

Gmelin knew that controlling such outbreaks would not be easy. Attempts to smoke beetles with sulfur or other substances in infested forests simply created villages of coughing peasants. "The bark beetle endures every kind of smoke, even if it reaches all the way to the maggot, whether it is straw, heather, sulfur, or arsenic smoke, without any damage." A mixture of potash and vitriol proved just as dangerous. Electricity didn't look very promising as an effective measure either. That left extensive logging as the best option.

Drawing on the advice of foresters, Gmelin recommended what would become the basic commandments of beetle control: keep the forest clean; regularly survey trees for the "vermin"; swiftly cut and salvage infested wood; remove windblown trees; and use trap trees to ensnare the flying worm in the spring. To the latter end, Gmelin proposed the felling of fifty to one hundred spruce trees full of sap in four or five different locations to lure flying beetles. Once the insects had copulated in the trap trees, he wrote, their "trunks are to be carefully made into coal."

Gmelin recognized that beetle outbreaks often ended without human intervention. Ants, wasps, earwigs, and birds commonly kept the beetle in check. "But to leave our forests to this hope and fate," he argued, "would be just as irresponsible as ignoring the threat of locust swarms or plague at the border, when taking precautions can save the blessings of our fields and the lives of our fellow-citizens."

Incredibly, Gmelin identified the same beetle control problems that later confounded western landowners in Alaska, British Columbia, Colorado, and New Mexico. Thinning the forest allowed the wind "to set in and uproot remaining standing trees." Chopping down all the infested wood often proved impossible, "because there are not enough workers to fall all this diseased wood." The sudden abundance of "worm dried wood" depressed lumber markets. Dead standing wood rotted so quickly that it lost its commercial value. Spotting an infested tree was also a damnable headache: "And how easy it is to miss a worm hole."

Nevertheless, Gmelin outlined a bold mission for foresters. By the 1900s the German insect journal *Entomologische Glatter* had referenced some 1,800 papers and studies on bark beetles. During the same period, the German gospel of managed forests and surgical strikes on scolytids crossed the Atlantic. Beetle mania arrived on this side of the ocean at a receptive moment. Both Canada and the United States had systematically mowed down most of their old-growth forests on the eastern seaboard and were belatedly considering novel ideas such as conservation. For two hundred years, the favored technique for clearing a forest among North American pioneers was girdling, or the "Yankee method," which involved ring barking a live tree, much like beetles do, and later burning the corpse.

The early American empire, built on the casual exploitation of wood, faced a crisis at the end of the nineteenth century. U.S. loggers were then felling 3.8 billion cubic feet of forest a year, or nearly two-thirds of the world's supply. Dwindling supplies of wood products and the prospect of a timber famine now threatened the economy. Under those conditions, the German idea of domesticating trees and regulating forests gained civic appeal. In the process, the bark beetle became a demonic obstacle to both conservation and scientific logging alike.

Andrew Delmar Hopkins, a charismatic farm boy from West Virginia, put the beetle on the map in the United States with a Gmelin-like thoroughness. The budding self-educated entomologist wrote several short treatises on controlling the "insect enemies" of the forest. Hired by the West Virginia University in 1890 to document local insect troublemakers for just a dollar a day, the energetic Hopkins came upon a patch of longleaf pine on White Top Mountain killed by bark beetles. For the next three years, he studied and documented the "widespread depredations" of *Dendroctonus frontalis*, the famous southern pine beetle. Hopkins suspected that the bug had ravaged southern pine forests since time immemorial. It "attacked healthy trees and girdled and killed them by excavating long winding burrows beneath the living bark on the main trucks of the trees." He recommended felling and burning beetle-filled timber. But Hopkins, who would receive an honorary university degree for his southern pine beetle work, didn't stop there.

In 1898, the United States created a Division of Forestry headed by Gifford Pinchot, America's first professional tree protector and conservationist. Pinchot, who had received his forest training in Europe, sought to protect reserves of trees from "reckless lumbering" for future generations and to end "the suicidal policy of forest destruction." Until that time, Americans, like Canadians, generally assumed their forests were "inexhaustible" and not worthy of a second thought, even as watersheds. One of the first reserves set aside by the new forestry division encompassed the Black Hills in the Dakotas, an area of 60,000 square miles. But there, unfortunately, an unknown insect enemy threatened to kill what the government wanted to conserve:

magnificent stands of "bull," or ponderosa, pine. Reports from
the region typically described clouds of beetles "that landed on
cabins like locusts."

Pinchot didn't have a forest entomologist at his disposal, so
he hired Hopkins to check out the "the pine destroying beetle
of the Black Hills." During a four-day trip to the region in 1902,
Hopkins, "a near genius" according to his peers, collected nearly
five thousand specimens of wood-chewing insects, including
the "Black Hills Beetle": the infamous mountain pine beetle. In
subsequent reports, Hopkins described the pine beetle's "stout
form," "broad head," dark color, and distinctive J-shaped gallery.
He calculated that beetle "depredations" had killed a billion feet
of timber in the Black Hills alone. Hopkins, eager to promote his
new trade, assured his superiors that "under proper management,
a forest can be protected at moderate expenditure." Like most
economic forecasters, he stretched the truth.

In reality, the Black Hills beetle would make a fool out of
Hopkins. At first, the entomologist tried to quell the outbreak
with trap trees, cutting down hundreds of green trees to attract
flying beetles. But keeping "experimental trees sacred from the
ruthless hand of the honest sawmill man" proved impossible, he
found. Enterprising loggers just carted them off. Moreover, the
trap trees appeared to have no impact on beetle numbers. Hopkins next proposed cutting and then transporting thousands of
logs, along with their beetle broods, out of the Black Hills by train.
But market demand for "blue lumber" killed that idea. The forest
managers also had a mandate to save green trees.

That left Hopkins with only two options: felling and burning
trees on the spot, or debarking attacked trees up to twenty-five
feet in height with an ungainly tool invented by a prospector.
(The tool, with its twenty-foot handle, did work, but it was expensive.) Another bark beetle entrepreneur offered to electrocute the
"timber bug" with "20 miles of copper wire with cotton insulation,"

zapping bugs tree by tree and using a horse as transport. Hopkins reasoned "that it would be too big a job and too expensive for the Government to encourage." But he lamented that it was "too bad we have to turn down such advanced ideas."

In the end, the forest service simply cut, felled, and burned more than 15,000 trees between 1906 and 1908. Their effort made little difference. The epidemic waned on its own. Hopkins did something very human, however: he declared victory over the bug once it was gone. Thereby began a tradition of professionals making dubious claims about their ability to stop beetle outbreaks. Eager to prove the efficacy of a tax-paid service, Hopkins wrote, "There is no doubt in my mind that this [felling and burning] was the primary cause of ending the trouble." He would later boast that the felling and burning of a thousand trees in Colorado Springs stemmed a similar infestation by the pine beetle: "The forces of the enemy have thus been sufficiently weakened to make their complete subjugation a comparatively easy matter." (In the last four years, the U.S. Forest Service has treated and removed more beetle kill from the Black Hills than from any other forest.)

And so began a long, ineffective, and expensive war against bark beetles throughout the North American West. In 1912, J.M. Swaine, a Cornell graduate, opened up a new military front as Canada's first Dominion entomologist in charge of forest insect investigations. In a report on forest conditions in British Columbia, then Canada's logging frontier and the source of Sitka spruce for WWI airplanes, Swaine lamented the injury "being done to valuable timber" by beetles. In particular, he identified the western pine beetle as "a serious enemy" to both the province's ponderosa pines and its irate lumber barons. He alarmingly documented one "red-topped" outbreak that stretched all the way from Vernon to Nicola, a distance of 250 miles. In some infested trees, he found as many as 1,500 to 2,000 pairs of beetles. Citing

Hopkins's contrived record of "successfully controlled serious outbreaks of destructive Bark-beetles," Swaine recommended two remedies: floating infested trees in water and cutting and debarking trees that supported beetle broods.

Swaine followed up his 1914 pamphlet with a lengthy bulletin devoted to the "most insidious enemies of our forests": Canadian bark beetles. After identifying the traits and habits of several tree killers, Swaine optimistically suggested that the logging of green "beetle trees" during the winter could stop an outbreak. Without much evidence, he also suggested that fires could "obtain terrific headway" in masses of beetle kill. Since fire running through dead ponderosa trees in southern B.C. could reduce the region to a "timberless waste for generations," he said, a war against beetles was a war against fire, too.

Throughout the 1920s, U.S. and Canadian foresters and other insect warriors routinely overrated their ability to end the tree-killing bedlam of bark beetles. Ralph Hopping, who had started collecting beetles at the age of nine, set the tone. Writing in the *Proceedings of the Entomological Society of British Columbia* in 1923, Hopping argued that forests provided not only the most valuable crop but the most revenue for government. Therefore, it was the duty of the Entomological Branch to "combat these epidemic infestations," which Hopping called the "most serious disturbance of Nature's equilibrium." One eight-year-long pine beetle epidemic, for example, had destroyed 200 million feet of lumber worth $7 million in revenue, he said.

Hopping admitted that the enemy's enormous numbers made the job a daunting one. He calculated that a California sugar pine 9 feet in diameter could house a million beetles. One yellow pine just 24 inches in diameter contained more than 6,000 of "the killing species of beetles," and more than 16,000 "insects of all kinds," 10,000 of which "were more or less injurious." Controlling incipient outbreaks and burning slash, he said, would keep the beetle in check and save forests for sawmills. "The tremendous loss caused to the forest in the past [would] not only be stopped," Hopping promised, but much of the fire risk would "cease to exist."

The growing economic war against timber bugs slowly eroded the public's attachment to beetles in both North America and Europe. During the nineteenth century, the kingdom of Coleoptera was a source of wonder in public circles. But as more "gallant bands of workers selected by the government" stalked the forest to preserve government coffers from "destructive pests," the fascination with beetles died. In 1924, one traditional beetle collector sadly noted this transformation. He appreciated "the importance of the economic side of entomology" but lamented the loss of any other attitude. "The question of dollars and cents looms large on the mental horizon of the modern man... I belong to a class (a vanishing class, it may be) who study the insect purely for the inherent interest in the creature itself."

The ascendant class fought bark beetles with zeal. In British Columbia, Hopping employed hundreds of men and many hundreds of thousands of tax dollars to spot infested ponderosa pines and then fell, peel, and burn them. Hector Allan Richmond, who later became one of Canada's most celebrated foresters, served as one of Hopping's spotters for beetle-attacked trees in the 1920s. After watching the government hack its way through the forest for a decade, trying to outrun one long beetle outbreak, Richmond declared the operation an exercise in futility: "The infestation and dying of timber spread at a rate far in excess of the control program." He later noted that the insect "continued on its destructive mission until practically all the old growth pines were killed, making way for young, vigorous pine in nature's never-ending rotation within the forest."

Few foresters, however, shared that enlightened view. Consider the determined war against the western pine beetle (Dendroctonus brevicomis), a near clone of the mountain pine beetle. This stout brownish beetle, which creates winding, snake-like galleries seventeen inches long, dines on ponderosa and Coulter pine in California and a few southwest states, such as Colorado. Indians, trappers, and ranchers had never worried too much about outbreaks of this pine eater, because a beetle boom simply removed aging trees and opened up the forest. But when loggers realized that they were competing with the beetle for virgin stands of mature pine, the insect suddenly morphed into "a problem of great importance."

By the early 1900s, *Dendroctonus brevicomis* had become the subject of numerous elaborate forest insect investigations. One study found that as many as 8,625 adult beetles could attack a tree and produce as progeny 31,740 emerging attackers. Another investigation found that attacking beetles produced a noise that resembled "a faint crackling sound like the crushing of dried leaves." Industry estimated that the beetle chewed up a billion board feet a year ("a tremendous loss"), enough wood to build 100,000 five-room houses and employ 57,000 people on an annual payroll of $18 million. Not surprisingly, crushing the "Pine Beetle Logging Company" became a government priority in forest reserves.

For scientists, defeating the Beetle Logging Company meant finding the best way to kill the most beetles at the lowest cost. It wasn't an easy endeavor. In 1960, entomologists J.M. Miller and F.P. Keen laid out the grisly particulars with both wit and rigor in their drolly titled *Biology and Control of the Western Pine Beetle*.

Throughout the western U.S., Miller and Keen reported, researchers had experimented with every beetle-murdering tool imaginable. First they published instructive manuals on "Fell-Peel-Burn." It took a three-man crew one hundred minutes to cut six hundred board feet of infested trees and then peel and burn the bark, the manual outlined. But when spotters missed infested trees, the beetles just kept on reproducing. Next, scientists tried burying "infested bark in soil." But the beetle broods still hatched underground and worked their way to freedom.

In Oregon in 1919, the Klamath-Lake Counties Forest Fire Association proposed electrocuting beetle-plugged trees. In the lab, scientists triumphantly shot 110 volts through beetles placed on a wet blotter. But when they tried zapping a tree with 11,000 volts for 15 minutes, only the broods near the contact point got fried. Alternating currents didn't work much better. Scientists at the University of California experimented with radio waves but concluded existing frequencies had "no practical value in killing bark beetle broods." In 1938, wardens in Yosemite National Park gassed beetles like WWI soldiers in the trenches. They covered unpeeled infested trees with a synthetic rubber blanket and then pumped it full of methyl bromide, a potent pesticide. Researchers quickly abandoned the tool, because "making a gas tight fit of the rubber blanket around an infested trunk on a rocky ground" was close to impossible.

Determined bug killers then injected infested trees with deadly chemical cocktails. Nicotine sulfate killed both the phloem and the tree. Results with copper sulfate were "entirely disappointing." Hydrocarbons came next. Researchers sprayed ponderosa pine with kerosene and then charred the bark, but the fire-scarring didn't kill all the beetle broods beneath. With the help of the U.S. Army, forestry scientists employed "goop," a putty-like material composed of magnesium and bitumen, to make incendiary bombs. The goop, which burns at 5,400 degrees Fahrenheit, effectively roasted beetle broods in felled logs. But supplies of goop dried up after WWII. The incendiary also required "unusual precautions to avoid spontaneous combustion."

In the 1930s, the Standard Oil Research Laboratory in Richmond, California, ever anxious to sell more oil, offered to produce special beetle-killing products. Researchers sprayed pines with creosote, naphthalene, and diesel oil in the lab. But results in the forest proved so variable that researchers ended up burning the oiled trees instead, and the thickness of ponderosa bark actually protected beetle larvae. Oil-based kills also cost more than peel and burn. After WWII, scientists tried DDT mixed with diesel, chlordane, and toxaphene, but the results proved "highly erratic and inconsistent."

Despite the apparent futility of massive beetle campaigns, killing beetles made foresters and politicians look as if they were making progress against forces larger than themselves. Between 1925 and 1932, entomologists waged relentless war against the mountain pine beetle in Oregon's Crater Lake National Park. Mature lodgepole pines populated the park, and the park superintendent feared that mountain pine beetles would leave the place a "windblown sandy desert." Scores of young men felled thousands of trees and then burned them. When the fire risk grew too great, the bug fighters exposed the bark of fallen trees to the midday sun to boil the beetles under the bark. But if the crew didn't carefully rotate the whole tree, two-thirds of the beetle brood would still hatch. A federal entomologist directing the operation in 1929 took a military approach: "The control forces have given the enemy repeated setbacks, but until recently the beetles on the southern front have had their forces strengthened by reinforcements from the north... The ultimate victory is now in sight." The following year the war effort consumed another $33,000 and another 50,000 trees, with little effect. Only a cold snap and the death of all susceptible trees ended the killing. The bug men claimed credit for the victory, but one park warden openly admitted defeat: "We have failed so miserably on this project that it has reacted very unfavorably on our work in the region." Over the years to come, park wardens continued to mock the expense and the hubris of these beetle campaigns, sometimes in song:

> The bugs they're killing the timber
> They've worked for many a year,
> But the Entomologists come and prophecy

That they'll quickly disappear.
Oh! They ain't gwine fly no more.
But how in the hell can the bug men tell
They ain't gwine fly no more.
The crews are cutting the timber
The crews are peeling the bark,
The bug men say the beetles they'll slay
And clean them from the park.

In 1952, the U.S. Forest Service spent millions of dollars to equip thousands of men with 5-gallon tanks full of fuel oil and DDT in Colorado's Routt National Forest. Wearing no protective gear, the attackers sprayed 45 feet up each infested tree. Yet "all that work didn't stop the beetle infestation." Only cold weather did. Around the same time, researchers sprayed so much lindane on beetle-infested trees in the Black Hills that the pines actually glistened in the sun. Nevertheless, beetle activity remained higher in the treated area than in the unsprayed forest. In Operation Pushover, foresters revved up heavy tractors and mowed down 1,800 infested lodgepoles in Utah's Wasatch National Forest then sprayed anything left standing with pesticides. On and on it went. In 1958, *Popular Mechanics* declared the bark beetle "one of the 20 most destructive insects" in the United States.

Yet after their exhaustive 1960 review of the experiments aimed at the western pine beetle alone, Keen and Miller concluded that "the killing of beetles, no matter by what method, has only a limited effect in reducing tree mortality. The trend of epidemics is only temporarily altered by direct control." In other words, trying to prevent a bark beetle from doing its anointed work in an aging forest is about as fruitful as trying to stop a flood or an avalanche.

Moreover, the two entomologists marveled at the beetle's good work in the forest: "The Western Pine beetle infestations were making periodic thinnings, which released trees from stagnation. The beetles attacked and killed the old susceptible veterans—suppressed, intermediate, codominant trees. The thinning made openings for regeneration and thus stimulated the growth of young trees. The result was an uneven aged forest composed of even aged groups of trees. Moreover the beetles were determining the natural rotation of ponderosa pine on different sites."

THE SOUTHERN pine beetle has a history as bloody as that of its western mountain cousin. *Dendroctonus frontalis* eats all sorts of pine in Central and North America, including longleaf pine, as well as plantations of loblolly pine. Between 1960 and 1990, the beetle dispatched nearly a billion dollars' worth of pine trees throughout the south. Foresters typically denounce the animal as "the most destructive insect pest of pine forests in 13 Southeastern States and in parts of Mexico and Central America." In North Carolina, Moravian settlers observed the scolytid's tree-killing "mischief" in the 1750s. Early Texas pioneers even employed the beetle to clear pasture, by chipping the bark off pines that stood in the way of progress and leaning infested branches against them.

Throughout the twentieth century, the southern pine beetle dismayed foresters with its periodic and unpredictable infestations: "It is either abundant, killing up to 50 percent of the stands of pine over large areas and killing out groups of pine here and there throughout the country, *or* so rare during the intervening years that it is difficult even to make collections." Like the eager professionals who went after the mountain pine beetle, southerners resorted to smothering, drowning, and fell, peel, and burn.

During the 1950s and 1960s, foresters drenched millions of pine trees with cancer makers such as benzene, hexachloride (BHC), and lindane. The chemicals cost up to ten dollars a tree. They didn't stop the epidemics, but they did kill just about every beneficial insect in the neighborhood. By the 1970s, foresters had realized that once an infestation covered a thousand acres, there wasn't much they could do except salvage a small percentage of infested trees or simply let nature take its course. Nevertheless, the war against the southern pine beetle created forest agencies with such belligerent names as the Expanded Southern Pine Beetle Research and Applications Program. That agency published a hefty book on the beetle, full of compelling insights: "The impact of the southern pine beetle is that it kills trees."

IN THE 1980s, Canadian researchers, including beetle expert Les Safranyik, took beetle killing up another creative level. Spurred by the mountain pine beetle outbreak in the Chilcotin that Allan Carroll had witnessed as a child, they initiated a set of novel beetle-killing experiments. On a military range, scientists wrapped a detonating cord containing C4, a plastic explosive, around infested ponderosa trees at four-, eight-, and twelve-inch intervals. The idea was to blow the beetles under the bark to smithereens. The explosions mangled all the adults, but "by the next day a high proportion of the beetles had revived and moved about normally." Safranyik quickly concluded that blowing up beetles was "time consuming." Or, as the droll Hungarian later observed, "Such an approach on a large scale would be questionable."

Another group tried freezing mountain pine beetles with CO_2 and liquid nitrogen. Romanian scientists had once tried pumping liquid nitrogen onto spruce beetles, with "promising results." But that's not what the Canadians found. After cutting down eight trees drilled by the mountain pine beetle, researchers applied the freezing gases with a "snowhorn" nozzle for thirty seconds. The treatment killed only half the beetles, and it cost a bundle: to freeze just one attacked tree (labor included) cost $323 for the nitrogen and $127 for the CO_2. By contrast, peel and burn ran to only $100 a tree.

Around the same time, scientists started to experiment with verbenone, a chemical isolated from the hindguts of female bark beetles. As noted, in the wild, the chemical signals other beetles that a particular tree is full. Used in gelatins and perfumes, verbenone costs about $1,400 a pound. Researchers such as John Borden initially reckoned that a one-tenth-ounce pouch of verbenone stapled to a group of trees would fend off beetle attacks and send the beetles out on exhausting forays. (The B.C. company Pherotech International even sells verbenone pouches under the trade name "No Vacancy Pack.") Foresters also tried dumping plastic and biodegradable flakes laminated with verbenone on small patches of forest containing highly valued trees.

But fooling Mother Nature has not been easy. Tests on verbenone's efficacy indicate that it remains one expensive ($150 an acre) and highly inconsistent product. Hot, dry summers evaporate the chemical quickly, forcing up both the dosage and the cost. One Idaho researcher found that verbenone pouches reduced tree mortality for five years but then lost their utility. Others found that treated plots showed higher attack rates in outbreak areas than did nonperfumed groups. Even when combining verbenone use with logging infested trees and insecticide treatments, the community of Lac Le Jeune experienced an "unfortunate outcome" of mass attacks on 175 trees. In 2009, Ken Gibson, a researcher with the U.S. Forest Service in Missoula, Montana, frankly summed up the verbenone record: "Suffice it to say, after 7–8 years of fairly rigorous testing—and not a minor amount of operational use[—]if someone were to ask any of us if verbenone protects host trees from MPB attack, we would have to honestly reply, 'Sometimes!' A 'silver bullet' verbenone is not!"

IN 2005, the Xerces Society for Invertebrate Conservation, a group dedicated to preserving insect diversity, reviewed one hundred years of the research on controlling bark beetles. The literature showed that professionals had written hundreds of papers about their inability to control the bugs. One humbled researcher suggested as early as 1978 "that doing nothing and letting infestations run their course may be a viable option." Salvage logging made things worse by removing "snags, parasites, and predators from the forest system." The overall effectiveness of removing infested green trees remained largely "unproven." Thinning appeared to be totally ineffective during an outbreak and made no difference in forests hammered by drought. Beetle-logged forests also supported fewer insects, mammals, and fungi and less overall diversity. "Rather than combat insects as pests, we should view their population swings as indicators of changing conditions in these forests," concluded the authors of the Xerces review.

Hector Richmond reached the same conclusion nearly one hundred years ago. In the 1920s, the Canadian government sent the budding forester on a reconnoitering trip through British Columbia and Alberta to spy on "the number one enemy": *Dendroctonus*. Everywhere his horse took him, Richmond discovered pine and spruce beetles chewing away as nonchalantly as Sunday picnickers at the park. On one hot afternoon, he even witnessed the mass attack of a large ponderosa pine located by his tent.

"It was a spectacular sight," he wrote. "The trunk of the tree was swarming with literally thousands upon thousands of adult beetles all actively engaged in chewing their way through the bark to the inner wood. The entire swarm had arrived within an hour. By dawn the following morning, the entire population had penetrated the bark and were no longer visible."

Richmond could hear the chirping of the beetles in the afternoon heat at a distance of ten feet from the tree. In his elegant 1983 memoir, *Forever Green*, the wise forester recounts his direct witnessing of many dismal, Brobdingnagian efforts to control the work of Lilliputians: clear-cutting, thinning, and pesticide spraying. These beetle wars always yielded the same eventual outcome: "In due time the beetle ran its course and disappeared after having eliminated practically all the old growth of ponderosa pine throughout the region. Today there stands in its place, young vigorous pine, exactly the way nature planned it in the first place."

In 1942, Charles Elton, the father of animal ecology, wrote a quirky book about explosive population booms in rodents triggered by bad agricultural practices or other human mistakes. Despite intervention through poison, fumigation, and prayers, the "insurgent subterranean activity" of mice could perforate a field as if it were a colander. After detailing startling and peculiar epidemics of mice, shrews, and lemmings, Elton concluded that "Authority" tended to respond in the same way to every rodent outbreak. In *Voles, Mice and Lemmings*, the ecologist laid out a familiar pattern: "Voles multiply. Destruction reigns. There is dismay, followed by outcry, and demands to Authority. Authority remembers its experts or appoints some: they ought to know. The experts advise a Cure. The Cure can be almost anything: golden mice, holy water from Mecca, a Government Commission, a culture of bacteria, poison, prayers denunciatory or tactful, a new god, a trap, a Pied Piper. The Cures have only one thing in common: with a little patience they always work. They have never been known entirely to fail. Likewise they have never been known to prevent the next outbreak. For the cycle of abundance and scarcity has a rhythm of its own, and the Cures are applied just when the plague of voles is going to abate through its own loss of momentum."

After recounting more astonishing stories about booms and busts of voles among Scottish sheep or the violent ebb and flow of lemming cycles in Scandinavia, Elton ended his lively book with a moral appeal to readers: "We stand on a near shore of an ocean larger than any that Columbus explored, in which we can at present discern only a few islands rising out of the mist. Let us hope that wise government will train navigators and equip them to explore more closely the Islands of Vole, Mouse and Lemming." This greater understanding would take account of "wildness and interest and beauty," what Elton called "the unstable fabric of the living cosmos."

Bark beetles respect this cosmos. Over the last decade, spruce monocultures have come under intense insect attack in Germany, Switzerland, Sweden, and the Czech and Slovak Republics. The scene would not have surprised Johann Friedrich Gmelin or Charles Elton. After monstrous windstorms or drought, *Ips typographus* descended on weak plantation forests in large bursts. In some cases, foresters engaged in unrivaled "sanitary logging" and then replanted the artificial forests with more Norway spruce, confirming Alfred Einstein's definition of insanity. But in central Europe, some scientists have astutely recognized that beetles are correcting fragility in a system created by hundreds of years of human ignorance. After monitoring the impact on diversity of a beetle outbreak in Šumava National Park, two Czech biologists concluded that the beetle was actually restoring the natural character of the place: "Homogeneous spruce stands with a lack of decaying wood due to past forest management seem to be changing due to the bark beetle outbreak to more open stands with a high amount of dead wood and an increasing proportion of indigenous broadleaved species such as rowan and beech. Thus, the bark beetle should be considered as a natural tool for restoration of a natural character of the mountain spruce forests which were altered by human activities in the past." Yet in the long war against the beetle, no "authority" has yet recognized *Dendroctonus* as the "cure."

FIVE

The Wake of the Beetle

. . . . .

*"We abuse land because we regard it as a
commodity belonging to us. When we see land as a community
to which we belong, we use it with love and respect."*

ALDO LEOPOLD, *A Sand County Almanac*

LIKE MOST people living in the interior of British Columbia,
Dave Jorgenson works in the wake of the forest industry's
response to the hurricane of the beetle. It is not a comfortable
situation. But as Jorgenson puts it, "Every day offers a new set of
disasters and challenges, and you have to adapt." On a February
afternoon in 2010, the lanky third-generation logger stood on a
clear-cut just west of Williams Lake, off the Horsefly road. Tolko
Industries, a Canadian timber baron, had just mowed down the
forest, ostensibly to take out some beetle kill before it became
"worthless." Any timber not "merchantable" or the right diameter
was left on the snowy ground. Before the outbreak, loggers rarely
left more than 15 percent of the wood in the forest. But after the
beetle, industry routinely junked 50 percent of the wood in waste
piles to be burned. An industry based on minimizing costs and
maximizing volume isn't particularly careful, explains Jorgenson.

Disgusted by the wasteful practice, Jorgenson convinced the company to build a snake-rail fence with the leftovers. A log fence is not only kinder to wildlife than barbed wire but lasts longer, he says. It also controls the movement of cattle in the bush. For Jorgenson, the whole sordid beetle affair has just shed light on an old and very familiar western story: "Greed and stupidity make a lethal cocktail, and this industry's been drinking doubles for a long time."

One of the biggest fallouts from the beetle disaster, he says, was industry's go-wild behavior. In 2001, about 60 percent of industry's timber cut was composed of pine. By 2005, at the height of the outbreak, the share of pine being logged had dropped to 40 percent in some regions. In fact, only a quarter of the forests attacked by the beetle were truly pine dominant. "It's pretty outrageous," says Jorgenson. "Many companies were just targeting non-pine stands."

Jorgenson isn't impressed with clear-cutting as the dominant, cookie-cutter solution for salvaging dead pine. An industrial clear-cut typically removes everything on the forest floor, including saplings as well as young spruce and fir. The gaping hole in the forest canopy also makes standing trees around the edges more vulnerable to blowdown. High winds in the region now rip through clear-cuts downing thousands of trees around the edge. Blowdown, of course, serves as a great nursery for bark beetles. "The bugs get in it, and you get another big red crop of trees. Every year it gets worse."

In 2009, B.C.'s Forest Practices Board revealed the ugly scale of clear-cuts in beetle country. The board noted that some exceeded 250,000 acres and were on an order of magnitude greater than the largest size ever recommended by the chief forester or provincial biodiversity plans. "We are concerned about these large and very large patches because they are a somewhat unintended and largely unforeseen consequence of salvage harvesting," the agency's report said. In an incredible finding, the

board concluded, "The ecological consequences of salvage harvesting on a spatial scale" had "no precedent globally." The board
had been forewarned. A 2001 review commissioned by the province on the costs and benefits of clear-cutting beetle kill had
argued, in simple terms, that a forest renewed by bark beetles
was a much smarter economic proposition than a monster clear-
cut designed by humans with forestry degrees. As Jorgenson can
attest, such studies went unread.

Selective logging of dead pine could have prevented much of
the industrial mayhem, but it would have cost industry a few dollars more. "All you see is green when I'm done," says Jorgenson,
who works only with small, custom-made equipment. "The pine
is gone, and it looks like nothing was going on." The remaining
spruce, fir, and balsam grow faster and fatter. But with harvesting
capacity already exceeding what the mills could handle, no big
company wanted to pay a little extra to log carefully, or to protect
the forest's ability to employ communities in the future. "It was
all about short-term fiber and get big or get out," Jorgenson says.

But big things like old-growth forests always eventually collapse. "These massive companies that require high-fiber flows
are going to find it hard to make a go of it," predicts Jorgenson.
"There could be some serious challenges down the pipe." A 2010
report by the International Wood Markets Group concluded that
"one of North America's largest natural environmental disasters"
would soon create significant shortfalls in fiber, chips, and sawdust shavings as companies run out of green trees. The report
calculated that 16 major sawmills could close and that United
States would experience lumber shortages. A similar report by
the Central 1 Credit Union estimated job losses as high as 20,000
by 2028 due to reduced timber harvests.

Dave Jorgenson regards the outbreak as a natural disturbance
driven by the rapid extinction of cold winters. He says there were
really four large epicenters, and they "all took off simultaneously."
The majority of the infestation started on timber leases owned by

timber barons, so "it was unstoppable." He remembers working in the bush and being surrounded by massive flights of beetles. "The sun was filtering through the trees, and it was like a London fog. It was surreal."

In 2007, the logger was invited to speak at a forestry conference in Grand Prairie, Alberta—Peace River country. Alberta, a petro-state with a Texas mindset, was then removing single infested pines by helicopter in mountain hot spots in a vain attempt to prevent the beetle's relentless march across the Rockies. Jorgenson told the Albertans that industry mostly regarded the beetle as a fantastic opportunity to obtain cheap wood. Logging corporations had little resilience and no interest in the long-term welfare of the forest or people. "You'll never control the beetle," Jorgenson informed his audience. "You'll run out of money before you control the beetle. You'll do what Nature lets you. You're not in control." He wasn't invited back.

Jorgenson thinks more forests should be managed by the communities that live in them rather than by corporations "situated in Vancouver or London or God knows where else." Unlike corporations and governments, communities have to live with the consequences of their decisions. "We are going to have to make peace with the fact that we have to make more with less. We're running out of stuff. We had it all and ruined it."

NOT FAR from Jorgenson's worksite, beetle kill spawned a new industry in Williams Lake: a wood pellet factory. For many forest workers, the epidemic was never about insects, bad public policy, or climate change. It was simply, as one engineer put it, "an industrial dead tree utilization issue of staggering proportions." Grinding dead trees into biomass energy seemed like a perfectly natural idea to Pinnacle Pellets, which is now the Pinnacle Renewable Energy Group. Vice president Leroy Reitsma says that climate change gave western Canada both a beetle plague and an incredible opportunity. By "cleaning up the result of the climate," British Columbia could produce 330 million tons of wood pellets "in support of healing the earth's climate."

Pinnacle Pellets, which combines mill waste with shredded beetle kill, grew from one 66,000-ton operation in 2004 to five plants that now turn out 825,000 tons of wood pellets a year. Ocean tankers ship mountains of pellets the size of rabbit food or vitamins to Sweden and the Netherlands, where they help co-fire coal power plants and thereby reduce greenhouse gas emissions. Rules that require European countries to generate 20 percent of their electricity and heat from renewable sources essentially drive the artificial trade. As the pellet salesmen say, "If you're in the wood pellet business, you're not in the forest products business. You're in the energy business." According to one estimate, a ton of wood pellets equals the energy contained in 3.69 barrels of oil.

But turning beetle kill into wood pellets is no easy task. Big trucks running on fossil fuels must transport beetle kill, or "bush grind," to the pellet plant. There, more machines empty a mix of kill and sawdust onto conveyor belts. Machines powered by electric motors then compress, hammer, and grind the material. Next, a tumbler fired by natural gas reduces the moisture content of the mixture from 35 percent to 8 percent. (A pellet factory that uses more moist bush grind than sawdust in its mix will also use more energy to dry the material.) Finally, electric motors run the material through sieve-like machines to make about five tons of smooth pellets an hour. It takes more diesel fuel to transport the pellets by train to the port of Vancouver, from which it's a nine-thousand-mile ocean voyage to Europe.

B.C.'s new export pellet industry also comes with a few economic drawbacks. Although a small-scale industry that served local markets could be sustainable, long-distance trade is subject to volatile shipping costs as well as unreliable supplies. Beetle kill doesn't last much longer than a decade unless sitting on dry ground. A 2006 federal study calculated that "it is not economical to build facilities that require substantial capital and long payoff periods specifically to use this supply of beetle-killed fibre given the lack of long-term feedstock." Moreover, the report concluded, "the full costs to access, harvest, transport and return a stand to production completely dissipate the delivered fibre value for energy." The solution might be to plant so-called marginal land with "fast growing plantation fibre." Both Chinese and European companies are currently invading "marginal lands" throughout Africa with beetle-like speed to grow biofuels.

Nonetheless, pellet companies continue to sprout up across the West. In Colorado, New Earth Pellets promises "green energy made here and used here." Given that 100,000 beetle-killed trees topple every day in that state, New Earth Pellets claims it is just "doing its part to make sure these trees continue to make a contribution to a healthy planet."

ABOUT 60 miles to the west of Pinnacle's 24/7 pellet factory, rancher Randy Saugstad deals with another form of beetle fallout: a watershed ruined by Tolko's ravenous sawmill in Williams Lake. In 2009, the forest service allowed a 12-mile-long cut block near Saugstad's 300-acre ranch, where he works 150 animals on horseback outside of Big Creek Provincial Park. "They call it beetle kill, but there was lots of green wood and spruce here," says the fifty-nine-year-old. He says the mill logged an entire creek to its headwaters with predictable consequences: flooding and drought. In the spring, Saugstad's operation now has too much water, and by late August he is dealing with out-and-out water shortages. He says the mill and the forestry service refused to do a proper hydrology study, and to protest the indiscriminate logging he blocked the access road for a day.

During the 1990s, Saugstad says, local ranchers, foresters, and guide outfitters put together a fifty-year land use plan for the region complete with details on the size of cut blocks and how to decommission logging roads. "Everyone walked out with a smile. But in the end it went into the garbage. The mill can go and do what they want. It borders on criminality." The scenario Saugstad paints is as hellish as the growing cut block near his ranch. "The government gives the trees away at twenty-five cents, and the mills lose money producing cheap wood nobody wants, and now the taxpayer will pay to fix the roads and the watershed around here. Isn't that insanity? The beetle is not a problem, but the logging is."

What Saugstad witnessed on his watershed will likely become a concern in beetle country across the West. Ruthless salvage logging compounded by beetle kill has dramatically changed the flow of water throughout British Columbia. Consider the Baker Creek watershed outside Quesnel, an area the astute and critical-minded forest hydrologist Younes Alila has studied for years. It covers six hundred square miles of forest and feeds the mighty Fraser River. Prior to 1970, loggers removed only 13 percent of the forest in small patches, with little effect on local water flow. But by 2006, after the mills had pulled out a third of the timber and pine beetles had attacked the remaining pine, the amount of water filling creeks in the spring had risen by 31 percent. Floods that used to come at twenty-year intervals will now strike every three years, warns Alila.

When the hydrologist looked at what would happen twenty years down the road, once 80 percent of the watershed has been logged, the results were grim. By 2017, peak stream flows could be 92 percent higher than normal. Moreover, the gentle, flat landscape in B.C.'s interior pine country will synchronize and intensify these large water pulses in the spring. Indiscriminate salvage logging combined with lots of dead trees could lead to frequent floods over a sixty-year period that will threaten river health, dam integrity, fishing grounds, and the people living downstream.

Most of this potential floodwater comes from the trees. Healthy lodgepoles make formidable water towers and can gobble up half the rain that falls in the forest. During spring, they can pump up to eight gallons of water an hour, only to sweat that moisture out through their needles into the atmosphere. A beetle-infested tree, of course, abruptly shuts downs the water pump, leaving the moisture in the ground. In a beetle-killed forest, about 60 percent of the extra water comes from expired water towers.

The rest of the surplus water comes from beetle-made changes to snow cover. In a forest of gray poles, snow that would normally rest on green branches and evaporate over the winter now falls to the ground. With no canopy to shade it, the snow melts faster and earlier in the spring. As a result, the water table rises higher in a beetle-attacked forest. Yet a dead tree surrounded by new growth "still has a hydrological function," says Alila. That's not true for a clear-cut. Given that the beetle has removed 1 billion water towers and that industrial salvagers have created enormous wastelands with little regard for water flow, many of British Columbia's salmon-rich watersheds will likely experience frequent and spectacular spring flooding in the years to come.

A good two-hour drive north of Randy Saugstad's ranch sits the pretty city of Quesnel. With more than one hundred companies devoted to forestry, Quesnel bills itself as the "Woodsmart City." Or it did. Once surrounded by a largely pine-dominated forest, Quesnel now belongs to the empire of the beetle and has paid for the short-sightedness of an industry dominated by sawmills. Claude Paquet, the owner of Clan Logging, sits dejectedly at his wooden desk in an industrial building in the northwest end of town. "There is no forest in the forest anymore," he says.

The soft-spoken sixty-three-year-old started out thirty-seven years ago, logging by horse, but says his work will soon come to an end. He used to employ fifty men who would cut and haul saw logs to the mills. Now he's down to a crew of fifteen guys hauling beetle kill or sawmill waste to be turned into wood pellets. "It might get worse yet. The effect on this town will be felt for a long time," he laments.

In the bush, Paquet has faced the beetle many times. While he was fixing the lights on his caterpillar one summer evening in 2005, a swarm enveloped him. The insects started to bite his neck and arms. "It was like a cloud of rice. We could hear them in the bark. It's a very weird noise." Paquet says the climate changed, then the beetle multiplied. "They fly with the wind." He has even found beetles still alive in the trees in November.

Like most British Columbians, Paquet is not impressed with how the government handled the outbreak. "A beetle is like a fire. You have to get out front of it, not behind it." He thinks way too much time was spent worrying about mill supply at the expense of planning for the future. "Their vision was not very large." But then again, he adds, "people forget they are the government."

The saddest development, says Paquet, has been the destruction of pine plantations. Like most loggers, Paquet took pride in his work and diligently replanted. Year after year, he would return to the sites to see how the young pines were growing. "But the beetles killed everything. The entire crop. Even trees six inches wide. Hundreds of thousands of trees. They are all dead. I can't tell you what the forest will be like in sixty or seventy years."

The changes that Paquet sees in both the logging industry and the forest are profound. The ground is wet all year around now, and after a rain you can't operate machinery in the bush for weeks. Some companies now leave as much as 80 percent of the beetle kill piled in the woods, because it is so cracked and rotten. The beetle, he says, is a wake-up call. "It's time to change and move on to other resources to make a living, but it's hard when you've done this all your life."

Paquet's business manager, Christy Kennedy, predicts that things will be even worse in five years when there are no dead pine left to log. But she is philosophical. "Maybe it will turn us into a society about needs instead of wants. Maybe it will teach people some fiscal responsibility."

Nearly 300 miles north of Quesnel, the remote logging community of Mackenzie has lost two thousand workers as one mill after another shuts down. Surrounded by cut blocks and dead lodgepoles, it's the sort of place that exports 2×4s but imports windowpanes. Even the bar has closed. Although the beetle affected only about 25 percent of the region's mixed forest, the industrial wake engulfed the community. At his hardware store, decorated with dead pines out front, Ernie Graham doesn't think the beetle has been a problem. But government beetle policy is another matter. "They flooded the market with wood. There was a surplus, which drove the prices down, and then came the recession. Industry in turn collapsed, and now we have job losses." He thinks the B.C. government's worst decision was lowering stumpage fees from $25 a cubic meter to 25 cents to encourage salvage logging. "Is that good management? Do you have to give it away?" Graham once employed seven people in his store but now is down to two. "It will never be the same as it was."

Kevin Neary, a Mackenzie-based entrepreneur who sells to pulp mills, agrees that the hubris is gone. His shop used to employ seven people. Now he's down to himself. "These guys were producing beetle wood to beat the band, and then the bottom fell out of the lumber market," he says. If nothing else, the beetle illustrated the folly of building a concentrated milling industry with a focus on high yield and volume instead of community-based industries that made value-added goods such as furniture. According to Neary, "We need to reboot the system with some entrepreneurs." All the federal and provincial programs just "threw money at the problem and hoped it would go away, and that doesn't work."

Perhaps the beetle's greatest impact has been in Vanderhoof, a mill town in the geographic center of British Columbia. Located on a vast tableland of almost pure lodgepole forest, the community boasts one of North America's largest saw mills. The Canfor Plateau mill, a giant industrial beetle, yearly transforms 70 million cubic feet of pine every year into 2×4s and 2×6s for the U.S. and Chinese markets. An American railway publicist founded the town in the early 1900s. Settled by Mormons, Mennonites, and Texans, Vanderhoof failed at farming but then boomed with the lumber industry, which still accounts for 40 percent of all jobs and made Vanderhoof the Republican heartland of Canada.

Annerose Georgeson, a fifty-two-year-old artist and former mill worker, suspects the beetle will shrink Vanderhoof's population just as it has shrunk the living forest. Just about everyone in the Georgeson family, children of Swiss pioneers, depends on the industry for a living. Annerose's husband, Frank, works at the Canfor mill and calls urban folks "cappuccino-sucking concrete dwellers." Her brothers Andre and Hans do contract work in the bush. Her brother Nick operates a skidder. A cousin works as a mill electrician, and her uncle cuts railway ties. Until her father died, he worked at the mill too. Her mother, eighty years old, can still wield a chain saw. All of Georgeson's five children have at one time or another worked in the mill, planted trees, or performed beetle probes in Banff National Park. "You just can't make a living farming here," she says.

In the late '90s, Georgeson's brothers warned her that the beetles were coming. "They said it was a big deal, and that industry was trying to keep on top of it, but every year it got worse." She heard about outbreaks to the south in 100 Mile House and felt sorry for that community, and then the beetles arrived in Vanderhoof. "You really don't get it until they come to your backyard." The pines turned red, and the beach at the lake grew black with beetle carcasses. Whenever anyone cut a tree, beetle larvae "would fall out like rice out of a bag."

All the trees around Georgeson's house died, including the thirty-year-olds. Just about every living pine at her parents' two-section farm became a gray lodgepole skeleton. "It was supposed to be our inheritance, but the beetles turned a woodlot into a pine nursery." Because windstorms brought the dead trees down like matchsticks, the family pulled off fifty trucks' worth of lumber. "We hardly got a cent for it. Everyone was crying. It was horrible." As soon as residents of Vanderhoof smelled smoke from a 25,000-acre wildfire in 2004, everyone pulled out their chain saws. Wind patterns in the valley have changed altogether.

For the time being, Vanderhoof, a community of 4,400, remains in beetle-boom mode as loggers haul truckloads of dead pine to the supermill to be made into spaghetti 2×4s. But the loss of so many trees has subverted memories and residents' sense of place. Georgeson says that she routinely loses her bearings in the denuded landscape. "People all over the place are disorientated. Everything looks so different. Prince George was full of pines. I used to go here and there looking for certain pine landmarks. Now I have to read the road signs."

The size and proliferation of cut blocks around the community also astounds Georgeson. Once industry got permission from government to salvage beetle kill, they turned the forest into a quilt of clear-cuts so giant that it can be seen from outer space. "The scale is just freaky," adds Georgeson. In 2004, the government tripled the district's annual allowable cut to 230 million cubic feet. But by 2013, the harvest will fall down to 35 million cubic feet of green trees. At that point, the Canadian Forest Service predicts, Vanderhoof and other forest-based communities will find that "the assumption of a perpetual and relatively stable supply of timber" will have to be offset "by the need to diversify." Climate change, adds the 2008 report, will bring higher costs, increased instability, "unexpected changes in forest health," and the new beetle buzzword, "uncertainty."

On a warm February afternoon, driving down the Kluskus Forest Road south of town, Georgeson and a passenger passed cut block after cut block. The remaining forest of dead trees extended for nearly 150 miles, all the way to the Kenney Dam, she said. Every lodgepole plantation 30 years or older was half red or dead. Georgeson thought the silhouettes of dead pines looked lacey and wraith-like. In the ghost forest where she stopped to walk, a lone woodpecker hunted beetle larvae. A gusting wind made the tops of the needleless trees sway and creak like nineteenth-century ships on the ocean.

After she'd cut the trees around her house, Georgeson attended a beetle talk sponsored by government at the local theater. When the evening was over, feeling that the bureaucrats had missed the emotional human toll taken by the beetle storm, she decided to do something. The world's largest insect outbreak had a message, and it needed repeating: "Stop. Pay attention. Look at this. Don't just walk past and not notice!" So she raised some money through the Omineca Beetle Action Coalition for an art exhibit and put out calls for beetle art. Thirty-three people responded with forty-four works of art. "It was way more than I expected," she says.

The extraordinary collection included poetry, paintings, and sculptures. Every contributor submitted a short statement on the outbreak. Susan Barton-Tait, a well-known Canadian artist, made white paper tubes out of trees that she had to cut down. She called it *Rebirth*. Judy Blattner, who'd often heard beetles landing on her metal roof, painted a brilliant green sapling surrounded by dead parent trees. After felling trees on their property for a year, Blattner and her husband had replanted twenty thousand birch, fir, and spruce. Ranchers Dagmar and William Norton made a stark sculpture called *Hugged to Death*. It's a life-sized cemetery with lodgepole stumps serving as tombstones boxed by a white picket fence. The Nortons, who operate a top-notch bed and breakfast, believe that tree huggers prevented the government from clear-cutting Tweedsmuir and thereby doomed the entire region to beetle mania. "Our anger is directed toward those groups whose Envy, Greed and Ignorance significantly aided in the death of this organism," their statement reads.

Children from Sowchea Elementary School in Fort St. James painted stark watercolors of dead trees with titles like *Homeless* and *Standing Dead Still*. Claire Kujundzic, a popular artist from Wells, B.C., painted a lonely caribou in a vanishing forest, saying in her accompanying statement that "Nature is giving us a very strong message about tampering with our environment." Annerose Georgeson contributed several ink photocopies of beetle galleries, which she called *Secret Language*. Their shapes reminded her of ancient script, she said. Tom Dean did an eye-catching acrylic painting of burning trees called *Trial by Fire*: "When nature becomes disease it must be cleansed and fire is nature's way of cleansing impurities."

Thousands of residents flocked to see Red and Blue Beetle Art, which traveled to nearly a dozen rural communities in 2008. Visitors found the works both uplifting and painful. Chief Colleen Erikson of the Saik'uz First Nation wondered why anyone would want to make art about something so ugly and horrible. To that kind of response, Georgeson says she often replied that New Deal photography in the U.S. during the Great Depression was horrifying yet necessary. Without it, people wouldn't know what had happened. "I felt someone had to document the human point of view. I just wanted to give people a chance to tell their story." In mill towns experiencing big lay-offs, workers "cried when they came to the show," Georgeson says. The exhibit never made it to a big Canadian city like Vancouver or Toronto. "They don't have lodgepoles there," she explains.

Georgeson says the beetle outbreak made her feel "betrayed by Nature" at first. She soon recognized that collapse, renewal, and beetles are part of the same story. "But I also feel guilty about how my lifestyle partly caused all this." She doesn't think people switching to energy-efficient lightbulbs will slow down global warming enough to save northern forests from beetle invasions.

At Georgeson's parents' farm, grass, wild roses, and blueberries have taken the place of pines. Cedars, too, are emerging from the swamp. "I'm hopeful, I always am." Life, she says, "is not always about us" but is "as precarious as a game of pick-up sticks."

Many aboriginal people in British Columbia see the beetle epidemic as a sign of things to come. In the eighteenth century, a prophet by the name of Boba lived among the Saik'uz people in lodgepole country. She predicted the coming of the railroad, government pensions, and other marvels. To this day, elders pass down her stories. One of Boba's prophecies, as told by Veronica George, appears to be a distinct warning about the perils of climate change.

"And then what we survive on will come to an end and weird things will start happening in the world," George quotes Boba as saying. "Then the water is going to get hot. No matter how big the lake is, the water will still get hot. The sun too will become hotter, and because of this the water will become hot, and however many fish there are, all will die. And the earth is going to dry up and nothing will be able to grow because the sun will be so hot."

"We don't know what is going to happen in the future," George says, "but this old person saw it. 'Then the animals will have nothing to feed on, because there will be nothing growing anymore.' The earth will dry up because it will get so hot. She said we should watch out for this."

In beetle country, people can now put the ashes of loved ones into urns made of blue-stained beetle wood, which marketers call "denim pine." In Prince George, Doug Plato "decided to take advantage of the crisis facing central B.C. by making some lemonade." The retired woodworker turns bug wood into wine boxes or urns that come in three different sizes. Plato, with only a thousand-square-foot working space, sticks to urns, though outfits in Penticton and Colorado make full beetle-wood caskets.

"Crematoriums are the majority of our customers," he says. It takes a lot of patience to work with dry beetle wood. "It's labor intensive and very hard on tooling. The wood is just unpredictable." He throws out about 5 percent of his boxes and urns because of fine cracks and stresses in the wood.

The government of Canada helped to fund the Denim Pine Marketing Association. "Out of the devastation" rises a "huge international marketing opportunity," proclaimed one politician. Throughout the West, companies now make flooring, tables, cabinets, "beetle beds," stools, benches, vanities, dog food trays, and log homes out of denim pine, "the wood that nature colours." In Colorado, some businesspeople have even established the Beetle Kill Trade Association to advocate for the "recycling" of dead or dying lodgepole pines: "Otherwise Colorado will be known as 'the state with the dead trees.'"

Beetle-wood memorabilia sells briskly in communities that have lost their pines. In Prince George, visitors and locals can pick up blue-stained bowls, jewelry boxes, and wooden trucks at several local art galleries. "It's nice to see something beautiful from such a disaster. The tea lights fly off the shelves," says Kim Brown, a saleswoman. Beetle-kill products also make popular wedding gifts. Advertisements for tables and benches made by Alpine Furniture Company in Colorado declare that the "beetle won," but "we can make good use of what is left."

IN B.C.'S Anahim Lake at the south end of Tweedsmuir Provincial Park, where everyone thinks the beetle hurricane began, Dave Neads and his wife, Rosemary, live in a post and beam cabin they built in the bush by a ridge called the Precipice. Neads, a geographer and environmental activist who moves as large as a farmer in the forest, has worked on the beetle issue for a decade. He remains dumbfounded by the prevailing idea that government and industry could somehow have controlled such a naturally ferocious event. "Can you imagine the thinking? That somehow industry could take over an area the size of Switzerland, and then go in and log and control the beetle? It's faulty thinking. You haven't got the machinery or the people. A forest is not a wheat field that you can put a fence around. It was a local manifestation of a global event. There are some things we cannot control." He offers one sentence of advice to people living in the wake of the beetle in Colorado and Wyoming: "Don't make things worse by logging the daylights out of your forest, because it will make the fall down worse."

The second thing that angered Neads was the idea that if the beetle kill wasn't logged, fire would race across the province's interior faster than a stock-market crash. "It was inaccurate and not supported by the science," he says. A green forest with lots of terpenes will easily support a raging crown fire, explains Neads. But once a beetle-killed pine loses its needles, "the fire risk is gone." When the trees fall twenty or thirty years later, the risk of ground fires will indeed increase. But there is no scientific evidence that shows beetle-killed forests spontaneously erupt into raging wildfires. During the Great Yellowstone Fire of 1994, stands of beetle-killed pines didn't crackle and burn as severely as living lodgepoles did, because they didn't have the needles or the branches to carry a fire across the top of the forest. "The idea that red forest will explode and that fire will march over the hill just isn't true," says Neads. "It was an outright lie." Drought, however, will turn any forest into a tinderbox. (Stephen Pyne, a world expert on fire, agrees. "There is no single or universal outcome from an insect infestation," he says.)

The other myth that really irked Neads was the idea that a dead forest posed such an onerous carbon burden that it had to be chopped down or turned into alternative energy such as wood pellets. "A beetle-altered forest is neither dead nor a carbon source," he says. According to Neads, when the waste associated with salvage logging generated a pellet industry, government and industry pretended they were doing the world a favor by removing carbon bombs from the landscape. But he doesn't think exporting a million tons of wood pellets, one of the world's poorest fuel sources, halfway around the world to generate electricity solves any problems at all. Moreover, the supply of dead wood is so temporary that the industry can't be sustained unless it starts to plant trees for use as fuel.

A lodgepole castle is composed of 45 to 50 percent carbon. As it breathes in carbon dioxide, the tree uses the substance to build itself. The majority of the carbon gets locked in the tree, and the roots bury about a third of it in the ground. On average, a one-acre stand of trees holds about 170 tons of carbon. Although younger trees store more carbon than older pines do, a lodgepole never stops saving. "Old trees offer a lower interest rate, but they'll still be taking up carbon," explains Art Fredeen, a plant physiologist at the University of Northern British Columbia. Even dead trees hold their carbon for a long time. "The carbon loss is amortized as the tree slowly rots over time."

A beetle-killed forest is not a carbon liability, says Fredeen. When he looked into the issue, the forest instead told a story of remarkable resilience. Fredeen and several of his colleagues measured the carbon fluxes from three beetle-killed forests and two salvaged clear-cuts. Within a year of being attacked, the beetle-killed stands, which suffered between 80 and 95 percent mortality, had become important carbon storehouses again. The dead trees allowed more sunlight to reach the understory, which in turn promoted vigorous young tree growth. "It's not as though beetle kill results in a carbon catastrophe," he explains. In contrast, the clear-cuts remained a stubborn source of carbon for ten years. CO_2 monitors showed a sixfold increase in carbon releases from clear-cuts compared to beetle kill left naturally to decay. "There is no green stuff on the clear-cuts."

To Fredeen, the findings have important implications across

the West. Wherever the beetle has gone to work, government has worried about carbon losses and industry has lobbied to transform "the wasted wood" into wood pellets. "We should not predicate the whole salvage operation of pine beetle on the greenhouse gas argument," says Fredeen. Rather than rushing into the forest to cut down trees, he proposes less management, not more. "We should leave the stands with lots of regeneration. Just leave them alone."

Although the governments of British Columbia and Canada eventually spent more than a billion dollars battling a beetle, the money mostly went to infrastructure projects. Despite heady talk about beetle-proofing the forest industry, the money did what politicians understand. It cut down beetle kill, repaired roads damaged by logging trucks, constructed new highways, and fixed up parks and municipal centers. Lots of federal money also went into prospecting in beetle-killed forests. The federal government noted that "the area affected by the mountain pine beetle is thought to host significant mineral and energy deposits." As many as three gold, copper, and molybdenum mines have already been proposed in beetle-riddled forests.

In Alberta, the pine beetle has defied, if not subverted, the government's "control, salvage and prevention" program. In 2009, another "staggering" flight from B.C. deposited beetles "farther east, north and south than ever before." If the outbreak ever reaches the scale of B.C.'s tsunami, Alberta could increase its mill capacity by 20 percent and still never catch up. Like B.C., the province has a Mountain Pine Beetle Advisory Committee that issues periodic reports. In 2007, the committee warned the Alberta government that the insect was here to stay and would likely have a "profound impact" on fifty northern communities. Moreover, the committee recommended that government encourage rural Albertans to "make necessary adaptations to enhance and improve their resilience." In an appendix to their report, the committee warned about "the crisis after the crisis," citing closed mills, flooded landscapes, and low timber prices. "Consideration must be given to whether or not the system itself returns to status quo and whether the standard assumptions still hold." The Alberta government never released the report to the public.

Most scientists now suspect that global warming will intensify aridity so strongly that lodgepoles will become an endangered species along the eastern foothills of the Rockies. Without much emotion, the Alberta Forest Genetic Resources Council reports that forests that evolved over thousands of years will likely disappear by 2060: "The anticipated change in climate may be too rapid and severe for successful adaptation by our current forest trees."

Even though *Dendroctonus ponderosae* hasn't invaded Saskatchewan or Manitoba yet, both governments have officially declared it "a pest." In 2009, the government of Canada, which has the worst record of any industrial government on climate change, examined the risk the pine beetle posed to jack pines across the boreal forest. Government scientists confirmed that "immigrants have successfully reproduced and are in the early stages of range expansion." Given the patchiness of jack pine stands, no one is predicting a pine beetle freight train to Newfoundland. However, the report speculates that a slow-burning boreal epidemic will reduce timber supply, refashion the forest in unpredictable ways, and create greater fire hazards. The word "uncertainty" appears continually in the report, like pitch holes. Warmer weather will likely propel the beetle eastward to Saskatchewan, and abundant and denser pine forests thereafter could further the insect's advance all the way to Canada's easternmost province. The report's authors didn't think the beetle could be stopped: "There are few examples where populations have been successfully suppressed." Government scientists expect "the dynamic behaviour of beetles on an invasion front" to remain both imperial and grand. But nobody really knows for sure.

To this day, just about every British Columbian has a comment about the beetle outbreak. John Borden, the chemical ecologist and verbenone salesman, thinks that industry and government got complacent. "All hell broke loose, and we never caught up." He now believes that any beetle outbreak should be treated like a spot fire and be stamped out. As soon as we see smoke, we put a crew on it, says Borden. "We're really good at suppressing fire. In fact, we're way better at playing Smokey the Bear than suppressing beetles. If we treated the beetles like fire and put every smoking infestation out, we wouldn't have stopped this outbreak. It would still have occurred. I mean, it was so vast. But we could have prolonged it. Instead of a ten-year outbreak, we would have had a thirty- or forty-year outbreak. We could manage it so we don't have catastrophic events like this." Borden admits that fire, beetles, and wind all play a role in renewing the forest. "But we want to be the stand-renewing factor."

Les Safranyik, the retired forester and MPB expert, offers a different lesson. "Ecosystems are finite," says the biologist who still finds it hard to believe that a creature so lumbering as a bark beetle can do so much damage. "We should never look at a landscape as being etched into stone. The destruction of a forest, when stands fall apart, is not a pretty sight to see but it is a part of life and succession. Time means nothing in ecology or biology. Life simply goes on. One stage will end and another will start."

At the University of Northern British Columbia, Kathy Lewis, a forest pathologist, says the beetle emergency has delivered one simple message: "Diversity, diversity, and diversity." She says that's the key to ecosystem stability. "We have to have diversity to adapt." Doing the same thing everywhere on the landscape is not a smart option. "I think we've got the message, but making changes is another thing altogether," she points out.

Lewis notes that climate change has increased the prevalence of other troublemakers in the forest, such as red band needle blight. The native fungus, a bane of pine plantations in the United States, infects lodgepole needles and rots them off. Cool weather has kept the fungus in check in northern British Columbia. But over the last decade, the fungus exploded in lodgepole plantations around Smithers. "It likes things warm and wetter," Lewis says. "It's been killing mature trees as well. And that's unheard of."

In 2010, an evangelical pastor drove from Alberta to B.C. to attend a wedding. Jeff Jones of Calgary's Calvary Grace Church could have been driving from Whitefish to Missoula, Montana, or from Fort Collins to Colorado Springs, Colorado, or from anywhere to anywhere else in the West. After seeing all the dead and red trees along Highway 5 between Merritt and Kamloops, Jones composed a sermon entitled "Theology of Impotence, or the Pine Beetle and the Fall." The pastor noted that one of the most prosperous places in the world had been reduced to praying for cold winters. It reminded him of Alexander the Great, who succumbed to a mosquito bite after conquering the Mediterranean and the Persian Empire. "The red forests of B.C. remind us that we are truly at God's mercy," Jones wrote. "It is God who declares, 'and all the trees of the field shall know that I am the Lord; I bring low the high tree, and make high the low tree, dry up the green tree and make the dry tree flourish.'"

PERHAPS NO western artist or woodsman has thought more about the meaning of the beetle than Peter von Tiesenhausen. The fifty-one-year-old is to landscape art what Ansel Adams was to photography. The son of Baltic German homesteaders, von Tiesenhausen now owns a quarter section of wooded land near Demmitt, Alberta, close to the British Columbia border in the Peace River country. He lives just a short drive from the place where the lodgepole forest of the eastern Rockies meets and melds into the jack pines that extend lazily across Canada as if on a conifer quilt. Von Tiesenhausen literally lives on the bridge that the beetle crossed to invade the boreal forest. He had a front-row seat to what the beetle Nostradamus Jesse Logan has called "a potential biogeographic event of continental scale with unknown but potentially devastating, ecological consequences implied by an invasive, native species."

Like the writer Cormac McCarthy, von Tiesenhausen doesn't try to explain his work. He lets his sculpture, paintings, etchings, and video start their own conversations, much like a startling landscape does. With a chain saw he often sculpts trees into stoic, muscular, larger-than-life-sized figures and then burns them. Several of these Watchers guard his property. "They're all looking up but have no eyes," he points out. "It's about standing firm and being awake and aware in the moment." He once took a crew of his Watchers right across Canada in an old pickup truck and through the Northwest Passage on a Coast Guard icebreaker. He also makes ships out of willow branches and puts odd-shaped baskets in trees. His art invites people to remember their connections to the land, which von Tiesenhausen feels should be as strong as a beetle's ties to a pine forest. He once famously kept an oil and gas company off his property by proclaiming that his art-laden forest was protected under Canadian copyright law.

As the pine forests grew red throughout B.C., von Tiesenhausen knew his land was next in line. He created a beetle-inspired work, called *Requiem*, for an exhibit at the Two Rivers Gallery in Prince George. Using local pine pulp as a canvas and ash from beetle-killed pines as paint, he drew five hundred silhouettes of lodgepoles. Each painting represented the memory or ghost of a tree. He etched panels of plywood with a chain saw and an ax to create images of fire. He charred strips of wood, and then etched small images of Watchers. From a distance, the entire piece looks like an aboriginal pictograph. The exhibit, which was displayed in a churchlike setting, so moved the pineless people of Prince George that the city later commissioned a sculpture from von Tiesenhausen to commemorate the beetle tsunami. A five-thousand-pound iron Watcher, originally carved from a dead pine, now guards the gallery's entrance. A twelve-foot-high bronze sculpture of a beetle-killed pine springs out of the Watcher's head, Athena-like.

Most of the pines on Tiesenhausen's property are dead now. He had 250 mature lodgepoles, but over three years beetle flights silenced them one by one. "These guys are just amazing. Some trees got hit so hard that they were solid pitch tubes," he says. In July 2010, "a massive flight" of beetles dropped out of a blue sky on the artist as he bucked wood in the bush. The insects were so hungry they bit the artist's neck and arm and attacked nearby aspens. Full of grief and helplessness, von Tiesenhausen set one of the afflicted one-hundred-foot lodgepoles on fire one winter and watched it burn like a giant, unworldly torch. To the artist it looked like a pestkreutz, a sculpture commissioned to appease the gods, ward off plagues, and honor the dead.

In 2008, von Tiesenhausen had a revelation. "When you come to terms with death and grief, you can see what to do with it," he says. He helped to organize a meeting of the Demmitt Cultural Society, which at the time had but four members and no cash. Today the society hosts major musical events and boasts 250 participants. A revitalized community has raised nearly a million dollars to build a straw-bale community center out of denim pine. It will be the greenest and most beautiful public building in a rural land scarred, like most of the West, by the careless extraction of natural gas. Much of the wood was horse-logged off von Tiesenhausen's own land.

The artist says the community-center project is the hardest thing he's ever undertaken. But the passing of so many pines and associated childhood memories forced him to ask a question few dare to ask anymore in the West. It's a sort of Gary Cooper or Clint Eastwood question: "What am I'm going to do with the time I have left?" The beetle gave von Tiesenhausen an answer: "I better make it meaningful."

Requiem for a Forest
David Jorgenson

There's a region in central B.C.,
Where the money it's said grew on trees.
But the pine trees turned red,
For the pine trees were dead,
And the problem is one of degrees.

Now the climate is warmer, you see,
Which is good for the bugs that eat trees.
So the beetles are fine,
And on pine trees they'll dine,
Turn the heat up a bit more, if you please.

Then the government made up a plan,
To deal with this blight on the land.

Let's cut all the dead pine trees down,
Haul the logs to the mills in the towns.
But the mills wanted green trees instead,
For they're worth more than trees which are dead.

So they clearcut the spruce and the fir,
Forests that are became forests that were.

The dead pine will be left to decay,
While the green trees are all hauled away,
Till the wood supply's down,
Cash and jobs then leave town,
And it's those who remain that will pay.

Soon a landscape once heavily treed,
From its arboreal cloak will be freed,
And the rubble that's left,
Of all trees now bereft,
Is a legacy to corporate greed.

SIX

The Ghost Forest

. . . . .

"Between every two pines is a doorway to a new world."

JOHN MUIR, *My First Summer in Sierra*

EVERY SUMMER, Louisa Willcox, a prominent grizzly bear advocate and avid mountain climber, hikes to the summit of Packsaddle Peak in Tom Miner Basin, Montana, like a determined mother on a mission. Although Willcox enjoys the expansive high-country vista into Yellowstone National Park, she mostly comes now to check on the state of her ancient children: the silvery-skinned whitebark pine. In early August 2009, the pines are not doing well, and Wilcox is growing anxious. "Things are getting ugly," she says.

The whitebark pine is a five-needled monarch that lives in majestic groves throughout the Rocky Mountains and the Sierra Nevada–Cascade ranges. It can be found in at least twenty-five U.S. national forests and much of the mountainous interior of British Columbia and Alberta. This hardy, long-living tree sports outrageously purple cones. The tree itself stores more carbon than a lodgepole. The whitebark pine can endure high winds, poor soils, Arctic cold, deep snow, and lightning strikes. The biologist Diana Tomback calls it "a tenacious survivor." But these gnarled and shrubby-looking pines are now dying by the millions, which could unravel another iconic part of the West in ways no one ever expected.

When Willcox visited the Packsaddle a month earlier, in July, "the ugliness" was not yet there. At the time, she was giving a tour of the forest to members of the Greater Yellowstone Whitebark Assessment Project. The unusual team consisted of retired beetle expert Jesse Logan, conservation pilot Bruce Gordon, and Wally Macfarlane, a veteran mountain man and landscape analyst. Together with the U.S. Forest Service, the team was about to map the scale of beetle damage in whitebark country with an unprecedented aerial survey of Greater Yellowstone. To that point, the government had not collected much comprehensive data on beetle attacks in the pines, because it considered the tree of little commercial value. Willcox, a long-time member of the Natural Resources Defense Council, thought that foolhardy. She even maxed out her credit card to get the map project going.

Using a simple classification scheme that ranged from zero to five ("no dead" to "all dead"), the team rated conditions on Packsaddle as a category two or three. The place was full of large, living trees. But from the summit, the team could see a "red cancerous expanse of dying forests" for hundreds of miles, and within a month what was once a vibrant forest on Packsaddle had become a land of zombie trees: they looked green but were deader than doornails. It was clear now that in a couple of months the mountain would be a category-four territory aflame with red needles. The deathly quiet of the forest had a British Columbian tenor, but with one startling difference. These trees weren't the abundant hundred-year-old lodgepoles but rare eight-hundred-year-old elders.

On her August ascent, which started in a spruce forest badly rusted and harassed by the budworm, Willcox passed monster whitebarks decorated with the dismal carpentry of the beetle. The garish, orange-colored pitch tubes on the trees resembled lurid, otherworldly tree blood. At the base of every pine lay fresh telltale mounds of light-colored frass, a combination of sawdust and beetle shit. As Willcox spotted one attacked tree after another, the wildlife advocate muttered in dismay. Some of the attacked trees predated Christopher Columbus's arrival. "It's unbelievable," said Willcox. "This tiny bug and climate change. It's no match."

The mountain pine beetle's assault on subalpine whitebark forests may seem at first a minor chapter in the great beetle Iliad. But the ancient whitebark anchors much of the West in ways few people appreciate. It not only serves as the crown for the Greater Yellowstone Ecosystem (GYE), home to the world's first national park, but guards the rooftop of a continent. Perched on some of the continent's most inhospitable and rocky terrain, these noble pine forests quietly regulate snowmelt, conserve water, shelter diversity, freshen the air, and grow fat-rich seeds for wildlife over an area the size of South Carolina in the GYE alone. The region shelters one of the most intact temperate forests on the planet and encompasses two national parks, six national forests, and twenty-one mountain ranges.

But in Yellowstone's high country and throughout the Rocky Mountain West, the beetle has now become what Jesse Logan calls "the fire that doesn't burn out." Since 2005, the epidemic has killed 7 million whitebark pines, creating "ghost forests," patches of gray skeletal sticks, on mountaintops from Oregon to Wyoming. Even the U.S. Forest Service admits that the mountain pine beetle, working in combination with a deadly fungal invader and climate change, has already placed whitebark and its hardy relatives such as limber pine "on the brink of disaster." For Willcox, Logan, and a host of other researchers, the whole mess unhappily proves the veracity of an old observation by John Muir: "When we try to pick out anything by itself, we find it's hitched to the rest of the universe." The whitebark appears to be hitched to the heart of hope itself.

FEW THINGS on this earth live as long as a whitebark pine. They are the grandmothers and grandfathers of the mountains. On OLDLIST, a database of ancient trees, *Pinus albicaulis* ranks 16th among tree elders, just behind the Douglas-fir. The oldest recorded whitebark lived over 1,200 years. Its venerable cousin, the bristlecone, often lives longer than 4,000 years and remains the oldest living tree on earth. Because of their age, five-needled pines remain the best record of climate change in the West. The width or narrowness of their rings tell stories of drought and deluge and heat and cold. The whitebark knows that sons and daughters of Europeans settled the West during some of the wettest weather in the last thousand years.

The whitebark, a sacred tree among the Blackfoot, occupies the highest elevation of any tree in the West. In spite of its remoteness, the generous tree glues together the alpine world and nourishes a remarkable community of animals and people. For thousands of years the Lillooet, Blackfoot, Shuswap, Kootenay, and Shoshone regularly harvested the pine's large, high-energy seeds, with their amazing fat, protein, and carbohydrate content. The pearly, pea-sized seeds also provide dinner for the Clark's nutcracker, red squirrels, and grizzly bears.

The tree's majesty enchanted naturalists such as the inveterate hiker John Muir. In Yosemite, he marveled at the rubbery suppleness of the pine's branches, which he could tie "like a whipcord." With the aid of a lens, Muir patiently counted 426 tree rings in a pine only 3 feet high and 6 inches in diameter. Under the branches of the whitebark, he found happiness. "During stormy nights I have often camped snugly beneath the interfacing arches of this little pine," he wrote. "The needles, which have accumulated for centuries, make fine beds, a fact well known to other mountaineers such as deer and wild sheep, who paw out oval hollows and lie beneath the larger trees in safe and comfortable concealment."

And the tree has many surprises. Unlike many conifers, which disperse their seeds by wind, the whitebark depends on the wings of the Clark's nutcracker. Bird and tree coevolved as an inseparable couple over a million years. As a backpacking biology student in the 1970s, Diana Tomback rested under an unfamiliar tree in the Sierra Nevada that was being visited by a strange black and gray bird with a long beak. When she asked a U.S. Forest Service wilderness ranger what the creature was, he said it was "a pine crow." Fascinated, Tomback started to study the unlikely pair, and she spent the next 30 years documenting how the two performed like an old married couple. Her research in the Sierras eventually revealed that one nutcracker could store 35,000 seeds in 9,500 caches in a year (two to three times the bird's own energy needs) and still recall their exact location. Wherever she watched a bird cache seeds in the ground, Tomback found newly germinated clusters of seedlings the next spring. "I was awestruck," she says. Incredibly, the nutcracker carries pine seeds in a special throat pouch, much the same way bark beetles carry blue-stain fungi. "The whitebark put its faith in the nutcracker, and the two became intertwined," explains Tomback, who now directs the Whitebark Pine Ecosystem Foundation, based in Missoula, Montana. "This tree is really a poster child for the mess things are in right now."

The tree's troubles began long before the beetle transgressed its normal boundaries and invaded the tops of mountains. Historically, groves of open stands of whitebark burned every fifty to three hundred years. The fire sparked by Indians or lightning guaranteed a diversity of pine ages (with few aging boomers) in high country and kept competitors such as alpine fir to a minimum. After a fire, seeds diligently planted by the

Clark's nutcracker would sprout to life. Fire, in short, allowed the pine to remain the monarch of high places. But after the Big Blowup of 1910, which consumed 3 million acres in Idaho, Montana, and Washington state and killed nearly a hundred people, the forest service got serious about suppressing fire. In the Flathead National Forest, for example, the agents of Smokey the Bear stamped out blazes and reduced the acreage burned from 2 million to fifty thousand acres in just twenty years. Fire suppression reduced the odds of a lightning strike or a spark rearranging a whitebark pine forest to once every three thousand years. (It did the same for lodgepole and ponderosa pine, creating the largest patina of beetle fodder ever assembled in the West.) As a consequence, the whitebark matured to become prime pine beetle food. Its abundance also diminished as spruce and fir multiplied and replaced whitebark groves.

Then along came white pine blister rust. It's a spunky Asian fungus with a bizarre, alien-like life cycle: it exists as five different kinds of spores that bounce between a variety of hosts, including gooseberries, Indian paintbrush, and whitebark. Once a spore invades a tree's stomate, or breathing hole, its fungal fingers slowly penetrate the tree's vascular tissue. After a few years a canker resembling an orange crescent erupts through the bark, killing stems and branches. Ultimately, the rust shuts down cone production and silences the tree. Although Asian white pine species can resist the fungus, their New World cousins, including sugar pine, bristlecone pine, limber pine, and whitebark, have little immunity to the Old World killer.

Deforestation and global trade, of course, brought the deadly fungal invader to the West. It's a tangled tale of one human mistake after another. During the 1880s, Europeans tried to restore their denuded forests by planting eastern white pine seedlings from the United States. The foreign pines grew well in European soils, but imported Siberian stone pines containing rust introduced the killer fungus into the plantations, and the fungus quickly undid all the restoration work. That dismal experience, unfortunately, didn't stop U.S. and Canadian west-coast nurseries from importing millions of French white pine seedlings in order to replant ferocious clear-cuts at the turn of the twentieth century. Not surprisingly, imported seedlings carried the unwanted fungi. It started a slow-burn epidemic that has changed western forests in ways that bark beetles never could.

By 1921, the rust had emerged in five-needled pines and currant berries in Vancouver's arboretum and throughout Washington state. Thanks to the region's wet, cool weather, the rust burst upon the landscape like a manic fire, with spores traveling up to three hundred miles on the wind. Idaho's glorious western white pine forests were the first ecosystem to go down. The Idaho pines, which once dominated the state's forests, provided jobs for five thousand loggers at the time. But the rust extirpated the trees. In an attempt to stop the fungal killer, the forest service employed twelve thousand men on blister rust control crews between 1930 and 1944. *Popular Mechanics* suggested that innovation and technology might eventually save the trees, but they didn't. Following vast lines of twine to identify rust-invaded areas, the crews dug up or sprayed infected currant and gooseberry bushes. But in the end, the combined destruction of salvage logging and rust turned western white pine into a relic.

The fungus also invaded whitebark forests, and by the 1940s most of the high alpine country in British Columbia and Montana's Glacier National Park had been infected. Forty years later, the rust could be found as far south as New Mexico and in every kind of five-needled pine, including the world's oldest organism, the bristlecone pine. Between the 1970s and 2009, the proportion of infected whitebark pine in Greater Yellowstone grew from 7 to 40 percent. "We knew the tree was on a trajectory that was not good," says Diana Tomback.

JUST AS white pine blister rust was starting to creep into Yellowstone, David Mattson, then a grizzly bear biologist with the national park, made some surprising discoveries about the importance of the pine seed in the bear's diet. Yellowstone grizzlies have a unique and broad-ranging menu that was known to include earthworms, hornets, cutthroat trout, army moths, dead · bison, and biscuit root. But after a colleague noted an odd correlation between the size of the whitebark cone crop and the number of conflicts between hunters and bears, Mattson explored the matter more deeply.

What he found wasn't all that surprising. During abundant seed years (every two to three years), healthy numbers of bruins congregated under the canopy of whitebarks to rob squirrel caches. The forest provided a sort of mountain refuge with few people around. During poor seed years, "the bears moved to low elevation near humans. They weren't starving. They were just on the landscape at large." That often put the animals into conflict with hunters.

Next, Mattson looked at the pine seed's impact on bear population growth rates. The sweet-tasting seeds appeared to be a fertility drug. During prolific crop years, grizzly numbers increased by 7 percent. After a lean seed harvest, the number of grizzlies dropped by 5 percent. Without pine seeds on the menu, bears also foraged farther afield, where they often encountered the primary agent of their destruction: people with guns and vehicles.

Mattson also examined the relationship between pine seeds and grizzly fecundity. The Yellowstone standard was two cubs, but Mattson found that a good seed crop made it more likely for female grizzlies to produce not only healthy cubs but a three-cub family to boot. Females ate twice as many of the nutritious seeds as males in late summer and fall, though males chowed down on twice as much meat. In the shade of the whitebarks, the females would smash the cones with their paws and lick the seeds with their tongues for hours on end. Females carefully extended the bounty by also looting squirrel middens the following spring. They could locate a cache under six feet of snow.

Mattson suspected that the bear's dependence on the tree might stem from an ancient time when grizzlies roamed as freely as the Blackfoot and the Lakota. "For bears, the tree is the energy and catalyst of their lives. It drives what happens in the course of their lives: where they are and what they are doing and even the health and number of their cubs." The only other place in the world where bears rely so heavily on the fat growing on pine trees is Siberia's Lake Baikal.

When it became apparent in the 1990s that climate change could dry up and dramatically shrink whitebark country, Mattson's research suddenly became controversial. The critical importance of pine seeds suggested that the only way to prevent extensive bear losses in Yellowstone was to change management policies and allow the animals greater room to forage for fattening foods in the fall. That would mean fostering more creative arrangements with hunters, ranchers, and gas companies, and such a notion left authorities flummoxed. For a long time, they choose to ignore the science and its implications. Partly because of his direct and repeated warnings, Mattson now studies cougars in Arizona.

WHEN THE mountain pine beetle rudely entered the picture in the early 1990s, the development grabbed the attention of forest researcher Jesse Logan, then head of the Forest Service's beetle research group in Utah. Logan, a wiry man who likes to ski and fish the high country, had been studying the effects of temperature on the life cycles of insects for years. Like many entomologists, he already knew that cold-blooded animals responded faster to climate change than did hot-blooded politicians.

And like a growing number of researchers, Logan also recognized that the mountain pine beetle was a formidable "agent of change" in the lodgepole forests. He figured that the lodgepole and the beetle have an understanding about death and renewal, "even if we don't fully understand it ourselves. They've worked out a deal." But he didn't think the beetle had worked out a similar arrangement with whitebark.

For most of the twentieth century, cold winters and cool summers kept beetles out of high-country pine castles. For the whitebark, the pine beetle was mostly a furtive visitor. But during the 1960s and 1970s, several severe beetle outbreaks turned great stands of whitebarks into ghost forests that remain skeletal to this day. The ferocity of the attack so stunned forestry officials of the time that they feared the mountain pine beetle could "boil over" into adjoining lodgepole pine stands. Some even proposed "the elimination of mature whitebark pine" as a beetle-harboring menace to lodgepole. But Logan looked at the history and wondered if the winds had carried the beetle into the trees or the beetles had set up shop on their own. Given the history of beetle outbreaks in warm spells, he thought the whitebarks needed serious watching. In the mid-1990s, he set up four monitoring stations to collect data on weather and phloem temperature high in the White Cloud Mountains of central Idaho at a place called Railroad Ridge. His superiors in Washington, D.C. criticized the initiative, asking him, "Why are you studying an insect that doesn't occur in an ecosystem that doesn't matter?"

Yet Logan persisted. He put together a computer model on how temperature might affect the pine beetle's behavior and life cycle with the help of mathematician Jim Powell and Canadian entomologist Jacques Régnière. The model suggested that just a 3.5-degree-Fahrenheit rise in temperature could propel the beetle to new heights. Higher temperatures both decreased beetle winter mortality and allowed the insects to emerge in unison to launch mass attacks, all in a one-year life cycle. Warmer winters and summers on a regular basis could turn the creatures into a scolytid version of Hannibal's alpine-hardened armies.

In 1994, at an insect conference in Hawaii, Logan raised the possibility that climate change might unleash beetle hell across the West. In a paper on climate change and beetle dynamics, Logan noted that insects routinely gamble during periods of climate change. They can sacrifice great numbers without going extinct. In particular, he warned that aggressive bark beetles might well overrun whitebark territory, cause massive die-offs, and potentially "foreclose evolutionary options" for the tree. Many colleagues advised Logan to spend his time on something that mattered. Today, many researchers call the retired scientist "the Beetle Nostradamus."

Logan watched his predictions come to life in high-elevation lodgepole country in central Idaho's Stanley Basin sometime around 1998. The mountains there normally cover the valleys with a cold-air blanket that keeps the pine beetle at bay except for the odd outbreak. (The basin's weather conditions strongly resemble those in Tweedsmuir Provincial Park in British Columbia.) As a result, the beetle had never gone on a boom-and-bust tear through cattle country. But after a cold winter in 1993 shut down an incipient beetle uprising, the insect bounced back and went on a record killing spree. As the weather warmed, most of the basin turned deathly red. Landowners tried to save their trees with thinning and verbenone traps, only to watch windstorms knock down the shallow-rooted survivors. By 2000, the beetle was reproducing within one year, just like its northern relatives in Alaska and B.C.

At that point, Logan flagged the pine beetle as a creature uniquely responsive to global warming. In 2001, he and James Powell published a remarkably prescient paper entitled "Ghost Forests, Global Warming and the Mountain Pine Beetle" in

American Entomologist. Citing unusual beetle activity in British Columbia, the authors warned that warming temperatures would shift the beetle's range into northern forests as well as mountain pines. If the beetle empire moved north, the insect would be in a position to invade jack pine territory, and from there the beetle would have full access to the boreal forest. "There is no apparent reason why a waterfall effect would not follow, spilling across the North American continent to jack pine in the Great Lakes region," the authors warned.

As if cued by Logan's computer models, the beetles fulfilled one prediction after another. In 2003, a massive flight of bugs "breeched the continental divide" at Pine Pass to invade Canada's boreal forest. That same year, the beetles overran Railroad Ridge. Logan had originally predicted that the ridge's whitebark forests might become an explosive beetle housing development by 2030. But like many climate modelers, he underestimated the speed of the warming and the beetle's hair-trigger response. The extent and intensity of the tree die-off floored him.

Together with his colleagues in British Columbia and Alaska, Logan realized that he was witnessing "spectacular outbreaks" of unprecedented size. Any one of the beetle events in British Columbia, Alaska, or Greater Yellowstone would have been unusual, but "their simultaneous occurrence is nothing short of remarkable," he wrote. The rapid invasion of whitebark country disturbed him most. Logan felt he was watching "the collapse of a great ecosystem in one of the most remote and inaccessible habitats on the planet" in real time. Moreover, most people had no idea it was happening.

Beetle behavior in the ancient pines also startled and confounded beetle watchers. *Dendroctonus ponderosae* made out on whitebark as though it were steak instead of hamburger. Females led deadlier attacks with fewer swarms and laid more eggs. Adults overwintered in pines only to emerge and attack trees again. Surprisingly, the MPB often left nearby lodgepole forests untouched in order to exclusively drill and dine on whitebarks. The beetles preferentially selected trees wounded by blister rust, which reduces a pine's moisture content, but they also targeted uninfected trees and small populations of whitebark pines that had resisted the rust. It was like attacking hope itself.

WHITEBARK PINE

Logan found that his work on "the ecological consequences of climate change" was deemed to be politically incorrect. The administration of George W. Bush, a dedicated oilman, did not want to believe that human beings could change the climate as assuredly as a beetle could change a forest or an oil spill would later change the Gulf of Mexico. A 2007 House of Representatives report confirmed that the Bush administration had conducted a systematic effort to "censor climate scientists by controlling their access to the press and editing testimony to Congress." Logan was one of many casualties. He could no longer talk to the media without a PR watchdog from the Forest Service overseeing the interview. But the scientist had no interest in being a puppet. While showing Michelle Nijhuis of *High Country News* extensive whitebark pine die-off in the spring of 2004, the athletic researcher outskied his PR handler in order to slip in a few frank comments about the impacts of climate change on forests and on why "the unexpected is to be expected." The Bush administration "was never receptive to science or the truth," says Logan today. "They wanted science to be the fantasy of how they saw the world."

In 2005, Logan tried to organize a major public conference on climate change and bark beetles. He wanted to raise public awareness by letting westerners know that an event as momentous as the melting of Greenland's ice sheets was happening in their own tree-filled backyards. He proposed inviting frontline beetle experts such as Alaska ecologist Ed Berg and B.C.'s beetle expert Allan Carroll. Logan thought Robert Redford's Sundance Resort might be a good venue. "The organizers became very interested. Science and art was a good way to tell the story. I was really excited," he says. But senior officials at the U.S. Forest Service didn't agree. After making one excuse after another, the service flatly told Logan that the conference was not going to happen. "And that's when I retired," he recalls. "This is bullshit." Logan and his wife moved to Emigrant, Montana, outside the park's boundaries, but he didn't give up on the whitebark.

By 2007, beetle kill in whitebark forests had reached some startling extremes. In one stand in Yellowstone, near Avalanche Peak, the beetles exterminated more than 162 trees per acre, or 92 percent of the trees greater than 5 inches in diameter. Yet when U.S. government authorities delisted the grizzly bear as an endangered species that year, they declared that only 16 percent of the forests in the GYE area contained any significant beetle die-off. Logan thought that the figures were misleading, if not dead wrong. As a man who spent more than 100 days of the year skiing in high country, he reckoned beetle mortality probably affected the majority of whitebark forests.

By 2008, Yellowstone grizzlies were feeling the rising toll on pines. After a really poor seed year, the Interagency Grizzly Bear Study Team (yes, there is a group with this ungainly name) reported the death of seventy-nine bears, 13 percent of the total population. The deaths surpassed "allowable limits" for both male and female bears. Hunters accounted for more than half of the human-caused deaths; according to the study team, hunters shot five females with a total of seven cubs in self-defense as the fat-seekers approached elk and other big-game kill. The catastrophic death toll was part of a trend that reflected the growing

loss of food sources in the region. Since 2000, female bears looking for sustenance had been shot later and later in the year. "The
bears were taking chances and losing," explains Louisa Willcox. After twenty long years of intense conservation work for
the grizzly, she calls the 2008 figure "infuriating, tragic, and
unforgivable."

The park hadn't seen such a great die-off since 1969. That
year, authorities closed all the park's garbage dumps and forced
food-habituated bears into a wave of human conflicts. The dump
closures resulted in a minimum of 229 bear deaths and crashed
the grizzly population to approximately 140 animals. With the
help of the Endangered Species Act in 1975 (which simply made
human behavior less lethal), the grizzly miraculously recovered
to about 500 to 600 animals. Yet despite warnings about the
growing decline in grizzly food sources, the government had still
declared the bear out of danger in 2007.

Following the grizzly wipeout in 2008, Jesse Logan teamed
up with Willcox. The whitebark admirers wanted to get a handle
on actual beetle devastation in the high country. To do so, they
hired pilot Bruce Gordon, the president of EcoFlight, and landscape analyst Wally Macfarlane to do an extensive aerial survey.
In a Cessna 210, Gordon and company flew nearly 4,500 miles of
transect lines (over mountains with names like Froze-to-Death,
Crazy Woman, the Beartooths, and Tempest) looking for green,
red, and gray trees. Except for a few high ranges with cooling glaciers, they found ghost forests all over the place. "Climate change
is here and right in our faces," says Macfarlane, who mapped and
photographed the beetle's growing empire. "You don't have to go
to the polar ice caps. All you gotta do is look up the mountain."

In 2010, the team released its snapshot of mountain pine
beetle mortality at high elevations in Greater Yellowstone. The
survey indicated that more than 50 percent of the high country
was dead and that 82 percent showed "significant mortality." The

worst die-offs had occurred in the Beaverhead-Deerlodge, Shoshone and Gallatin national forests. Islands of living whitebarks appeared in the Tetons, the Beartooth Plateau, and the Wind River Range, places with lots of snow, cold winters, and glaciers. But as the report noted, the ice is melting. The study also concluded that "outbreaks are spreading much faster than the models predicted and the consequences appear more severe than simulations indicated."

The report pointed out that the death of pine in 2,500 drainage basins over an area the size of South Carolina was not good news. Without stands of the ancient pine acting as a fence, the snowpack will grow smaller. And without the protective shading of the pine, the snow will melt earlier and prompt more spring flooding. And without the pines to ration the release of cold water, the mountains will deliver less coolness to the rivers and streams below in late summer. As a consequence, the waters of the Yellowstone and Madison rivers will get too hot to sustain cold-loving native cutthroat trout. Without the hermit pine, the West will rediscover the meaning of aridity, and the change will unhitch things we don't yet understand. The tree's impact on water, the report said, will "far outweigh its physical presence on the landscape."

THE COLLAPSE of the whitebark pine community in Greater Yellowstone is not an isolated event. In west-central B.C., pine beetles and rust have killed 85 percent of low-elevation whitebark forests. In Alberta, where officials are not fond of wildlife or trees, the government acknowledges "a gloomy picture for the future of the species." Beetles have also begun to perforate the high-country oyamel fir forests of central Mexico. *Pseudohylesinus variegates* normally attacks trees stressed by root disease, drought, and air pollution. Stephen L. Wood, the Mormon taxonomist, said in his book that this beetle attacks "large fallen cut and

unthrifty trees," but its numerous cousins can kill overmature
trees as well. In response, authorities are doing the usual salvage
logging. Yet this sacred tree provides critical winter shelter for
the monarch butterfly. Every year, the insects fly two thousand
miles to pollinate plants and wildflowers throughout the West,
including Yellowstone. By now, the oyamel fir forest, the subject
of relentless abuse as well as logging, has lost as many community members as has the whitebark forest. According to Lincoln
Brower, a biologist who has studied monarchs for most of his life,
"the migration of the monarch buttery is an endangered natural
phenomenon. It could go down the drain." The beetles, as usual,
are merely drawing attention to great wounds already inflicted
by human tree killers.

All across the West, old-growth forests are failing. A team of
researchers led by Philip van Mantgem and Nathan Stephenson
at the U.S. Geological Survey recently looked at the health of
58,000 pines and firs in 76 old-growth forest plots in six western states and British Columbia. Most of the trees were more
than 200 years old and had been monitored since the 1950s. In
the last 25 years, the scientists found, tree mortality has doubled
while new growth has stagnated. Stephenson compared the situation to a town in which the death of grandparents and parents
exceeded childbirths by two to one. "If you saw that going on in
your home town, you'd be concerned," he told the *Washington Post*.

The two researchers concluded that global warming had
reduced the snowpack, quickened spring melt, and lengthened
summer droughts. The aridity, in turn, invited more insects and
disease. Stephenson compared tree death rates to interest on a
bank account, because "the effects compound over time. A doubling of death rates eventually could reduce average tree age in
a forest by half, thus reducing average tree size." The trend could
mean sparser forests and larger or more abrupt die-offs in the
future. But nobody really knows for sure. (In contrast, eastern

hardwood forests appear to be growing faster as a result of warming temperatures.)

Perhaps the most striking and rapid change has taken place among stands of quaking aspens. Since 2005, Colorado has lost nearly a quarter of its 3 million acres of aspen forest. The slender trees die so abruptly, with no signs of regeneration, that scientists call the condition "sudden aspen death," or SAD. Most of the die-off has taken place on south-facing slopes among female trees at the lower ranges of the tree's elevation (5,000 to 11,000 feet) in Utah, Arizona, Wyoming, and Colorado. In most of the dying trees, scientists have found two tiny bark beetles (*Trypophloeus populi* and *Procryphalus mucronatus*) so rare that most aspen experts had never heard of them. (Stephen L. Wood, however, had duly recorded their existence as attackers of the weak, stressed, and dying.)

Most forest pathologists suspect that the aspens are going under because of acute drought. Scientists say the die-off is unlike anything they have ever seen. Small groves of 150-year-old aspen routinely die out, but now entire landscapes are succumbing. Like the mountain-loving whitebark, the aspen is an arboreal elder. It regenerates as cloned suckers from a single parent that can weigh 6,600 tons underground. One Utah clone that may be 8,000 years old covers 106 acres and has 47,000 individual trees. But with SAD the stems die first. Next goes the mature tree. Finally, the entire clone, one of the world's oldest living organisms, gives up the ghost. Forest experts now expect the range of aspen, one of North America's most common trees, to shrink by 40 percent throughout the West by 2030. The town of Aspen, Colorado, may have to consider changing its name as the brilliant yellow fall colors become a "big brown hole." A 2009 study sadly concluded that aspen "may be a prime indicator of the impacts of a changing climate on forest growth and productivity."

Many Asian relatives of the whitebark pine also seem to be in trouble. Suitable mountain homes for Japanese stone pine could decline by 25 percent and disappear altogether from southern mountain ranges. The nut harvest of stone pines in northern Spain has declined from 160 pounds an acre to just 90 pounds in the last 40 years, due to hot and dry weather.

AS THE beetle has signaled, the whitebark pine's future is decidedly uncertain. Despite valiant efforts to plant rust-resistant seedlings and thin out encroaching spruce with fire in high places, Diana Tomback and others worry about the tree's resilience. Fewer Clark's nutcrackers visit groves of whitebark decimated by rust or beetles, but the pine has no other careful planter and groomer. Tomback fears that the near extirpation of whitebark in much of Glacier National Park could foretell the tree's future in Yellowstone. "Whether we are capable of getting enough resistant seedlings in the ground and restoring a big enough area is really a good question," she says.

If the tree goes, grizzly bears in whitebark country will likely follow in a long decline. In Flagstaff, Arizona, Dave Mattson has switched from grizzlies to study the fate of cougars and other carnivores in a drying and fragmented West. The celebrated biologist calls the whitebark a magical tree. "In the fall, when the low light hits the trees and you see blue sky and hear the chattering of squirrels and the call of the nutcrackers, and you can smell the freshly dug middens, it's pretty damn compelling. For me, I grieve for the loss of this species."

In the 1970s, Mattson worked briefly on a beetle crew in the Black Hills, his birthplace, fighting *Dendroctonus ponderosae*. "My job was to mark attacked trees with paint so other members of the crew could cut down the attacked tree before the beetle spread." He says he laughed at the effort even then. A bark beetle attacking a lodgepole, says Mattson, is as wild a creature as a male grizzly pouncing on a rutting moose. "We are really reaching our limits, and creatures like the bark beetle challenge the imposition of our desires and wishes on this world." By defying our illusion of power, the beetle naturally "creates a crisis," and in response, people "demonize the agent of nature that defies our wishes and turn it into a problem." Bark beetles, Mattson believes, "are one of the agents that require us to revise our sense of ourselves in this world and understand our place on this earth. That's why I kind of like them, for the same reason I like grizzly bears. They both challenge our sense of who we are."

When Mattson looks at the data on climate change and considers how it will affect the movement of wildlife in a land as arid as the West, he shakes his head. It's foolish to think that we can mitigate and manage the effects of lost glaciers, ghost forests, and disappearing groundwater. "We can't fathom the changes about to face us or their magnitude," he says, any more than a mature whitebark forest could anticipate a swarm of beetles.

When people ask Jesse Logan about stopping the beetle, he informs them it's a natural event on the scale of Katrina. "Could you build a fan big enough to blow a hurricane back out to the ocean?" Until people burn fewer fossil fuels, harvest fewer trees, and plow less land, he says, "what is going to happen is going to happen."

But the death of so many whitebark pines saddens him. "This is a tree that has nurtured and sustained life for thousands of years. It has provided refuge for elk. It has provided food for the red squirrel and the grizzly. It has protected snow so trout don't stress out in the summer waters." The tree has seen everything that could happen in the high country, he points out. "Lightning. Beetles. Fires. Climate cooling and climate warming. It has nurtured a whole web of life and has that sense of being the Old Man of the Mountain. It has tremendous knowledge. Like a grandfather."

In the summer of 2009, Logan nursed a punctured lung and a scar above his right eye from a late-spring skiing accident. The scientist who had predicted the politically unpalatable sat in a chair before his computer with a photograph of a giant white-bark on the screen. When asked what the pine might feel about its current predicament, he searches for an answer. Scientists are trained not to think about trees as sentient beings.

"What might the white bark pine feel?" he ponders. "Betrayed."

SEVEN

The Song of the Beetle

. . . . .

"All things move in music and write it."
JOHN MUIR, letter to Jeanne C. Carr, Yosemite

DURING 2005, David Dunn often wandered the hilly outskirts of Santa Fe looking like a medieval plague doctor. Armed with headphones and a tape recorder, the avant-garde music composer and violin player poked the thin bark of pinyon trees with a special homemade device. The odd contraption consisted of a meat thermometer and a piezoelectric transducer from a Hallmark greeting card. After inserting the modified thermometer-cum-microphone into the tree's inner bark, Dunn patiently listened to the voices inside the tree. The bespectacled artist made an ungainly apparition in the desert forest as he perched against trees for hours on end.

Dunn became a tree whisperer after New Mexico started to lose half of its famed pinyon trees to an unprecedented beetle outbreak. Anxious landowners wanted a clear diagnosis on their trees before they pulled out their chain saws. Because Dunn had the listening tools, he got recruited for the job. Whenever the sound engineer heard noises that resembled running water or creaking winds in a pinyon, he'd give the tree an all-clear for beetles. Such a diagnosis inevitably invited two possible prescriptions: the landowner could water the tree more often, to build resin resistance, or he or she could spray the pinyon with the pesticide carbaryl. If Dunn heard squirrel-like pops and clicks, that meant the beetle had taken up residence and was now building its own magical sound universe. Such a diagnosis invariably resulted in someone pulling out a saw. When people offered to pay for his unique service, Dunn gracefully accepted a donation on behalf of his nonprofit Art and Science Laboratory. Dunn, after all, was collecting data on one of the world's most remarkable animals for one of the strangest and most unlikely of science experiments.

Dunn's otherworldly investigation began in summer of 2003, when clouds of *Ips confusus* alighted on the pinyons. The beetles turned the trees a deathly ocher, and then a Halloween gray. The musician's neighbors started to panic, but Dunn got curious and cobbled together his odd-looking microphone. "I felt there was an exceptional amount of biological activity going on, and I wondered if there would be any sound in the tree," he recalls. He discovered a richer acoustical world in pinyons invaded by polygamous beetles than he could have ever imagined. It was like encountering a skilled percussion group in the middle of the Mojave Desert. All the clicks and pops prompted the illustrious artist to ask a series of questions that scientists rarely ask.

Dunn started with the stories of Pueblo elders, who believe that "the beetles come when the trees cry." He wondered if that was true, and, if so, how a pinyon might weep. He was also curious about how bark beetles communicated in their winding galleries. Why did scientists know so little about the insect's acoustic abilities? Could the death of beetle-riddled spruce trees in Alaska be related to the pinyon-killing drought in New Mexico? His list kept growing.

Ips confusus, the innocuous pinyon engraver, set Dunn off on a seven-year investigation into scolytids, climate, beetle music, and complexity. In his spare time, Dunn diagnosed infested trees. The art world "needs to ground people's imaginations in a deeper understanding of the natural world," he believes. "I think it is essential at this point that artists take a role in collaboration with the scientific world—that artists and scientists work together towards real-world problem solving. We need all the help we can get."

Dunn's reflections produced some radical conclusions. Given the insect's evolutionary success and its ability to change entire landscapes at the drop of a hat, Dunn thinks bark beetles might be one of the most important animals on earth. "They are amazing creatures. They eat themselves out of a food source. That's a terrifying proposition." After producing a highly unusual beetle CD called *The Sound of Light in Trees*, Dunn began a wildly inventive collaboration to test an innovative idea: acoustic warfare against beetles. The results could change the entire field of pest management. "We altered beetle behavior by playing back their own sound," explains Dunn. "We managed to turn them into cannibals. We created unprecedented behaviors."

DUNN'S BEETLE odyssey took place in one of the West's most ancient and important woodlands. A pinyon juniper forest may look like a bunch of stubby trees on a model train set, but it defines the soul of the southwest. About six species of the aromatic juniper shade eleven different kinds of pinyon pine in these semi-desert lands. Most common among the pines are the Colorado pinyon (*Pinus edulis*) and the singleleaf pinyon (*Pinus monophylla*). A juniper, which has the longest roots of any tree, can live for 1,200 years. A pinyon can grow up to 800 years old. The random cowboy marriage of pinyon and juniper can take place in a variety of landscapes, including dense forests, shrub lands, and open savannas.

Hardy pinyon juniper woodlands now cover about 89 million acres of dry and sandy tablelands throughout California, Utah, Arizona, Colorado, and New Mexico. Sandwiched between desert and high alpine groves of ponderosa pines, these old forests shrink in hot weather but expand during cool, rainy times. The apple-shaped pinyon not only provides shade for coyote and antelope but protects critical watersheds on mountain slopes. About one thousand different plants, insects, birds, and mammals live in a pinyon juniper forest. The pinyon, however, is one of the slowest-growing pines on the planet. It takes sixty years to reach a height of six feet and about a century before the tree can produce a good crop of nut-laden cones. Edward Abbey, the famous forest ranger and writer, often rested in the shade of a pinyon tree just "to admire the splendor of the landscape and the perfection of the silence."

For nearly ten thousand years, the Hopi, Shoshone, Pueblo, Numu, and Navajo well understood the worth of the pinyon tree. The nut gatherers basically managed *Pinus edulis* and *Pinus monophylla* forests as nut orchards. They thinned and fired the forest to ensure mixed-age stands of high-producing nut pines, and for good reason. Rich in protein, fat, iron, and vitamin D, the pinyon nut is a civilization builder and a culture maker. Ronald Lanner, the U.S. naturalist and former forest researcher, notes that the Colorado pinyon nut provides more calories than a "pound of chocolate and nearly as much as in a pound of butter." The nut also contains more essential amino acids than cornmeal does. Bushels of the nuts have sustained desert communities in times of scarcity. (Abbey thought the pinyon nut was as sweet as a hazelnut.) The so-called desert manna makes not only a long-lasting flour but fine soups, stews, and porridges. After the rainy season, the abundance of pinyon nuts can energize the forest in a magical way. Following a good nut year, explains one pinyon gatherer, "you should see the mice and then the coyotes. Then the raptors. It comes in a huge pulse, and it all goes back to the pinyon."

Gathering pinyon nuts in the fall was a joyful and celebrated event. John Muir witnessed the energy and splendor of a nut-gathering expedition by Mono Lake in 1870.

> When the crop is ripe, the Indians make ready the long beating-poles; bags, baskets, mats, and sacks are collected; the women out at service among the settlers, washing or drudging, assemble at the family huts; the men leave their ranch work; old and young, all are mounted on ponies and start in great glee to the nut-lands, forming curiously picturesque cavalcades; flaming scarfs and calico skirts stream loosely over the knotty ponies... Arriving at some well-known central point where grass and water are found, the squaws with baskets, the men with poles ascend the ridges to the laden trees, followed by the children. Then the beating begins right merrily, the burs fly in every direction, rolling down the slopes, lodging here and there against rocks and sage-bushes, chased and gathered by the women and children with fine natural gladness. Smoke-columns speedily mark the joyful scene of their labors as the roasting fires are kindled, and, at night, assembled in gay circles garrulous as jays, they begin the first nut feast of the season.

For hundreds of years, many tribes traded the famine-fighting protein across the West and into Mexico. Until the 1930s, millions of pounds of nuts went to markets in New York and Los Angeles. Today, thanks to globalization, most Americans are stuck buying foul-tasting pinyon nuts from China, Korea, or Russia.

White settlers didn't appreciate the cultural significance or the pulse of the forest. Most pioneers singularly valued the short pine as fuel or fence post. For nearly two hundred years, mining companies clear-cut miles of mature pinyons to make charcoal for silver smelters. For most of the last fifty years, ranchers looked upon the aggressive juniper and the scrubby pinyon "as weeds in need of eradication." As a consequence, land managers knocked down more than a million acres of pinyons with cables and chains tied between tractors. Since the 1980s, strict fire management has increased the density of surviving stands and allowed wealthy Californians to invade the forest with monster estates and gated communities.

PINYON

That's how bark beetles found the pinyon juniper forest in 2002: abused, fragmented, unappreciated, crowded, and dried out. The Southwest boasts the largest diversity of tree engineers in North America. As many as ten different species of bark beetles typically manage pinyon, ponderosa, and lodgepole pines throughout the region. They include *Ips confusus* (the pinyon engraver), the western pine beetle, the southern pine beetle, the mountain pine beetle, the round-headed pine beetle, and several twig beetles. Where the western and southern pine beetle and several *Ips* species led the charge against millions of ponderosa pines in 2002 in Arizona and Colorado, the *Ips*, accompanied by two twig eaters, exploded among the pinyons of New Mexico.

Ips confusus doesn't normally go on wild killing sprees. Tellingly, Stephen L. Wood gives the creature scant mention in his scolytid bible. Unlike the predatory mountain pine beetle, the *Ips* is an opportunist that takes out individually stressed or injured pinyons. The male leads the attack and then is joined by four or five females along with the usual crew of mites, nematodes, and various strains of fungi, including the classic blue-stainers. *Ips* larvae primarily feed on the phloem of the tree and carve long-winding galleries under the bark.

A good drought, however, threw both trees and beetles off balance. The dry spell descended on the Southwest in the mid-1990s, after twenty years of wet weather, and didn't let go until 2005. On average, a pinyon tree needs nine to fifteen inches of rainfall a year to thrive. The drought reduced that intake to next to nothing. Compared to a previous drought, from 1953 to 1956, this one delivered hotter weather. The soil water content got so low that pinyon trees stopped transpiring and photosynthesizing altogether. At the Los Alamos National Laboratory, ecologist David Breshears could look out his window at a pinyon study plot the size of a football field and actually watch the trees die. "I would see the trees go from vibrant green to pale, gasping green to pale brown to dropping all their needles," he remembers. Between 2002 and 2003, the die-off was so extensive that patches of graying trees could be seen from outer space. It extended over 4,600 square miles, an area the size of Germany's Black Forest.

Drought-stressed pinyons gave the engraver beetle an opportunity to run riot in a landscape of undefended castles. The mild-mannered Ips normally reproduces two generations a year. During the drought, some populations reproduced as many as five times annually. "There was an exponential explosion," explains David Dunn. In 2003 alone, the outburst reached an Alaskan scale, covering 800,000 acres in four states. At the time, local ecologists called it a perfect storm of tree death: "excessive foliage and biomass produced during the wet years of 1976–1995; a severe and prolonged drought that limits their capacity to resist normal insect attacks; and tree-killing native insects whose populations are growing exponentially as a result of unusually warm weather."

Like all bark beetle uprisings, the Ips revolution caught everyone by surprise. In 2002, the beetles browned the pinyons on the Sangre de Cristo Mountains outside of Santa Fe and wiped out 90 percent of the pinyons around Los Alamos. The city of Santa Fe hastily cobbled together a "pinyon initiative," declaring, "It's hard to imagine that an insect the size of a grain of rice could become public enemy number one in Santa Fe, but this is how many have felt about *Ips confusus*, the piñon bark beetle." The creature predictably generated heated debates about falling property values, pesticides, the cost of dead tree removal, and the threat of wildfire.

As in Alaska and British Columbia, authorities generally exaggerated the risks of wildfire as well as the benefits of cutting dead trees. An unusual public letter released by a group of forest ecologists clarified matters. The ecologists estimated that the fire risk would remain high for only a year, until all the trees' needles fell off. Thereafter, "future fire hazard even in the most heavily affected areas will likely be no greater than before the mortality event, and probably will be reduced because of the substantial decline in canopy fuels." In contrast, they said, cutting down dead trees would accelerate soil erosion and encourage the invasion of cheat grass, an Asian fire-loving invader that saps the western landscapes of all vigor and moisture.

The *Ips* storm was unlike anything entomologists had seen before. Mike Wagner, a jovial, ruddy-faced researcher at Northern Arizona University, found engravers as thick as Canadian blackflies just ten miles outside of Flagstaff. "I was on a field trip in the forest, and every time I opened my mouth a bark beetle would fly into it," says Wagner. It was like an insect hurricane blowing through the mesa. Hundreds of beetles would bore into the boles of the pinyon, and thousands of their offspring would emerge shortly afterward to attack new trees. The populations grew so dense that not a living pinyon of any age remains. In the space of two years, drought and the beetle converted a pinyon juniper forest into a juniper forest. And the junipers also got hit: the twig beetle, *Phloeosinus*, mauled the tops of many surviving trees. "What astounded me was the speed with which the beetle converted the forest," Wagner remembers.

By the time the beetles approached David Dunn's home in Santa Fe, they had finished off 55 million drought-stressed trees, 10 percent of the state's forests. Although Dunn knew little about bark beetles or their acoustic life then, he already understood that the natural world was startlingly musical. In 1990, he had released a series of underwater recordings taken from insect activity in ponds. The creatures, including a variety of water beetles, made wild tropical-like rasps and sputters. "Their alien variety seems unprecedented as if controlled by a mysterious but urgent logic," wrote Dunn in his liner notes. He called his CD *Chaos and the Emergent Mind of the Pond.*

As Dunn's *Chaos* CD amplified, insects do a lot more talking than scientists initially thought. With the aid of sensitive digital technology, researchers can now hear insects rubbing together a variety of body parts (an act called stridulation) to make a host of unique sounds. A recording of two thousand eastern subterranean female termite workers, for example, sounds a lot like a dentist working on a tooth cavity. The sounds made by palm weevil larvae call up the image of a dyslexic Morse code operator.

At first, Dunn had thought that bark beetles were a largely mute group. But then the musician came across a series of interesting studies on mountain pine beetles done in Oregon in the 1970s by Joseph Rudinsky. Rudinsky found that courting males struck up an aggressive chirp to attract mates or to ward off rivals. They did so by rubbing a sound organ on the back of their heads, called a pars striden, against the main part of their bodies. (The pars striden works like a gyro or percussion instrument with ridges on it.) Rudinsky had also noted that acoustic signals could set off chemical cues, and following that the focus of beetle research shifted to these chemical perfumes. Given that beetles don't fly at night (and therefore don't have to fear bats), researchers thought it highly unlikely that bark beetles possessed any ultrasound receptors. To Dunn, though, it appeared that scientists had opened a door on the acoustic ecology of bark beetles and then covered their ears. According to recent research, "the diversity of sound-producing organs in beetles is amazing and unmatched by any other order of insect."

Dunn had the idea of retrofitting a ten-dollar meat thermometer to act as a microphone, with the hollowed shaft serving as a wave guide. The piezobender disc, taken from a greeting card, acted as loudspeaker. Dunn, who now exports this beetle-listening device to Chinese researchers, first planted the tool in the phloem layer of a pinyon tree by his home in early spring of 2003. He recorded silence. The tree creaked in the wind, but that was it. After temperatures got warmer, he waited two weeks and then tried again. This time, he heard stirrings, pops, chirps, and clicks. The phloem and cambium layers of a pinyon tree "are an amazingly effective medium for acoustic communication," he says. Over the next two years, Dunn made recordings at hundreds of pinyon trees. He was able to do something few beetle academics have done: listen unhurriedly to the deliberate and chatty work of scolytids in the wild.

Some days Dunn inserted his recorder near nuptial tunnels, other times near pitch tubes. He discovered that stridulation "can go on continuously for days and weeks," long after the beetles have mated and excavated their tunnels. All of this talking suggested to Dunn that the beetle not only had "a more sophisticated organization than previously suspected" but also used perfume trails for long-distance communication in the forest, then switched to stridulation for short-distance chats inside trees. In fact, he discovered, the beetles created an endless chatter in young and old trees at almost every time of the year except winter. "The beetles weren't even quiet at the end of summer."

Dunn carefully noted the condition of the trees he tested, recording the color of their needles and the number of pitch tubes on the trunk. On severely stressed trees, the beetles bored through bark without encountering any resin resistance at all, noted Dunn. Trying to save an attacked tree seemed futile, since "it doesn't take long for a beetle to take a tree down."

Dunn discovered during his investigations that pinyon trees several hundred years old appeared to have an entirely different relationship with the beetle than did younger trees. Most of the elders had survived previous beetle visitations and had some sort of resilience. Some of their branches thrived though others seemed dead. But big concentrations of beetles still settled in these old trees. "Their behavior [there] was entirely different, as though they had entered an equilibrium state. They hadn't killed the tree but had established themselves. It was though something else was going on," Dunn recalls.

The trees made their own sad music. A parched pinyon typically produces a variety of powerful pops that sound like distant drumbeats. Botanists call this collapse of cells "cavitation" and use it as a measure for drought stress. Healthy trees pump water from their roots to needles via a series of pipes in the xylem. The system works under negative pressure. During a drought, air bubbles form in this vascular system, creating a sort of tree thrombosis. The cells eventually collapse with a popping sound. "They can implode with such tremendous instantaneous force that, under laboratory conditions, they have been measured to produce temperatures up to five thousand degrees centigrade," says Dunn. These cavitation events release both light and sound signals that fall between an audible range (twenty kilohertz) and an ultrasonic range (two thousand kilohertz). The human ear can hear frequencies up to twenty kilohertz, but many insects can detect up to two hundred kilohertz of sound. Dunn and a growing number of scientists suspect that bark beetles may home in on these cavitation signals. As the Pueblo put it, the beetles hear a tree crying in an arid land.

The Sound of Light in Trees, the CD Dunn made from his pinyon-tree recordings, showers listeners with bizarre beetle sounds.

Alien clicks and rude tweets seemingly rise above an orchestral background of drying wood. The stridulations of the I*ps* beetle in the castle of a pinyon sound alternatively subterranean and watery. Many of the squeaks and clicks call to mind an old man sitting in a rocking chair in another room, cleaning his briar pipe with a shank brush.

ABOUT THE same time as David Dunn was making his beetle recordings, Reagan McGuire, a fifty-six-year-old truck driver, pool hustler, and genuine character from Pittsburgh, read an article about bark beetles killing 74 million trees in Arizona and New Mexico. "I thought that was a lot and wanted to do something about it. I'm a tree hugger. I love trees," McGuire says. He recalled how the U.S. military had blasted Panama dictator Manuel Noriega out of his refuge in the Vatican embassy by playing heavy rock music day and night. (The blasting also drove the Vatican crazy.) Similar long-range acoustic devices were used to deter Somali pirates. McGuire, a freethinker, wondered if he could create the same kind of acoustic stress in a bark beetle. He called up Richard Hofstetter, the young beetle specialist at the University of Northern Arizona in Flagstaff. Hofstetter listened during a meeting with McGuire, though "he looked at me as though I was crazy," McGuire recalls. McGuire kept going back. Eventually Hofstetter, who studies the mites and fungi on the beetle bus, found some money and put McGuire to work in a lab at the university's forestry building, which is surrounded by ponderosa pines. Given the dismal history of bark beetle control, Hofstetter reckoned it wouldn't hurt to try something completely different.

While doing background research, McGuire came across Dunn's *Sound of Light in Trees* on the Internet and shared it with Hofstetter. The incredible range of chirping sounds convinced Hofstetter that the bark beetle had a much more complex way of communicating than previously thought and that McGuire wasn't crazy after all. McGuire visited Dunn for a week to learn how to record beetle sounds in trees. The two often talked through the night about beetles, music, aridity, and life. Since then, the entomologist, the musician, and the pool hustler have collaborated on one of the craziest science experiments in insect history.

In his lab, Hofstetter fashioned a "phloem sandwich" so that McGuire could view his beetle subjects while filming and recording their reactions to different sounds. This beetle version of an ant farm consists of two quarter-inch-thick plates of Plexiglas with a piece of sugar-rich phloem inserted in between. A single hole allows a beetle to enter its favorite realm. Although the phloem dries out quickly, the sandwich gives researchers a clear view of beetles doing their thing. The first time Hofstetter used a sandwich, he watched mites eat a bunch of beetle eggs.

To start off the experiments, McGuire played the most abrasive sounds he could think of: heavy metal and, amusingly, angry monologues by Rush Limbaugh, the infamous talk-radio host. "I wanted an authoritative, agitating, and repeatable voice I could play back again and again. I also wanted to stress the hell out of the beetles, and I thought that hate radio would do it," McGuire explains. (Hofstetter, a gentle man of Canadian ancestry, later posted a disclaimer on his website explaining that the use of Rush Limbaugh's voice "was not a political statement.")

But the beetles in the phloem sandwich ignored Limbaugh's bombast. The beetles didn't react, either, when McGuire played the man's voice backward. "They're smart critters," adds McGuire, who basically learned the essentials of bioacoustics and entomology on the job. (At first, he couldn't believe how delicate the beetles were: "They are really amazing creatures. They are dogs of war, yet very fragile. I don't hate them. I just want to kill them.") The beetles also ignored head-banging tunes by Metallica, as well as Guns N' Roses' "Welcome to the Jungle."

Hofstetter then proposed that the team try playing more beetle-like sounds. That's when things got really interesting.

(*Ips* appear to have as many as five distinct calls that range from one click to a series of Morse code–like chirps.) They first put a female pine beetle in the sandwich and then introduced a male to the entry hole. The male promptly stridulated with a chirp sound saying "I'm here. I'm here." Then McGuire played the sound of another male chirping. The female promptly abandoned the real male in a vain attempt to find the louder but virtual source. "In one case, she tunneled to the speaker and waited. We changed their reproductive behavior completely," says Hofstetter.

Another experiment highlighted the power of bark-beetle voices in a grim way. Both Hofstetter and Dunn wondered why there were no hybrid beetles in Arizona, given that the western and the southern pine beetle often attack the same ponderosa tree. To find out, they parked a female western bark beetle in a phloem sandwich, then introduced a male southern pine beetle. "The female started signaling by making weak pulsing sounds. The male moved towards her and started to make a terrifying loud stridulation sound. The female froze in her tracks. Then the male came up to her and chewed her in half lengthwise. It was sonic warfare," says Dunn.

After McGuire had recorded and modified the squirrel-like screech of the male southern pine beetle (a heart-rending sound), he played it back to a pair of western pine beetles. He watched a male mate with a female two or three times and then suddenly chew her to pieces. "You know, you don't see that in nature. That's not natural." In another study, a desperate southern pine beetle male chewed his way through the Plexiglas. He continued for two weeks after McGuire stopped playing the sound. "His mandibles were reduced to stubs," says McGuire. "It was extraordinary."

Sitting in a narrow room full of dried-up beetle sandwiches and piezo recorders, McGuire patiently documented which of the modified sounds drove beetles most crazy. He manipulated beetle voices to disrupt mating, tunneling, and reproduction. He even watched six male and six female *Ips* beetles tunnel in circles and avoid sex altogether. Patents are now pending on some of the sound recordings. A beetle audio device the size of a smartphone that can be fitted onto a tree is being field-tested. "If we can reduce the fitness or fecundity of beetles by even a fraction, it will be helpful," says Hofstetter. When the team released some preliminary findings in 2010, the media cranked out four hundred stories with headlines such as "Not Everyone Likes the Beetles' Sound" and "Bark Beetles Rocked by Sound." Many articles highlighted how useless Limbaugh's voice had been in the beetle fight.

If beetle sounds can defeat scolytids and temper their forest-eating behaviors, Hofstetter's team could change the course of entomology. Instead of poisoning termites, ants, cockroaches, and other pests, people might limit these tidy empires with sonic fences or disrupt their mating practices with horrific insect yells and shrieks. Protecting stored grain supplies could become as simple as flipping a switch. "Acoustic ecology could change the way we do pest control," admits Hofstetter.

AFTER THE great pinyon die-off, the New Mexico government published a report in 2005 on how climate change will affect the state. The report concluded that there would be less snow, regular water shortages, more beetles, fewer trees, and greater uncertainty. The paper included an unintentionally droll paragraph on the hazards of rapid change:

> Surprises are inevitable. New evidence from paleoclimatic records now show that climate changes and ecosystem responses are not always gradual, but can occur abruptly over a few decades or less. Complex human and natural systems often respond in a nonlinear manner to increasing stress. That is, they change gradually or not at all until a threshold ("tipping point") is reached, and then they change dramatically. Positive feedbacks can amplify the impacts of small

changes into enormous effects, such as when a wildfire grows slowly until it begins creating its own winds and "blows up" catastrophically.

The massive scale of tree death so impressed researchers at the University of Arizona that they dug up twenty mature pinyons and replanted them in Biosphere 2. It's a futuristic glass-enclosed facility, a sort of mini-earth, where scientists can study coral reefs, mangrove wetlands, and savanna grasslands under controlled conditions. The researchers put half the trees in an enclosure set at normal New Mexico room temperatures. The rest of the trees were placed in a room where the thermostat was jacked 7 degrees Fahrenheit above normal. These trees got a climate-change warming shock, which scientists predict will be the West's new weather of the future. Then scientists stopped watering both sets of trees.

The results of the experiment explained a few things about the great pinyon mortality. Drought killed the trees in the climate-warming room 28 percent faster than the pinyons exposed to normal temperatures. The trees stopped breathing and died. Given the region's hundred-year drought record, the scientists projected, that meant pinyon die-off would occur five times more frequently due to hotter weather. "I don't want to be alarmist, but that is a super-conservative projection, because the result relates only to temperature, not to increased drought or bark beetles, which we know will exacerbate the problem," explains David Breshears. "So we could be in for a lot more change." With beetles in the picture, tree mortality could actually double.

David Dunn believes that bark beetles illustrate the evolutionary range of human experience, from drum-beating hunter gatherers to orchestral farming civilizations. He views *Dendroctonus* as a fairly primitive step in beetle evolution, with only one or two sound-generating organs. "They are monogamous hunter-predators. They take down a tree." *Dendroctonus* are such committed predators, however, that they can use the chemical signals of a tree as a weapon against itself. And the *Ips*, he muses, are much more sophisticated. "These so-called secondary species have an entirely different mating strategy and a different relationship with fungus." Because *Ips* are more interested in farming than in killing, they communicate more fluidly in a tree. They play different organs on their bodies to make sounds. Last come the ambrosia beetles. One clan peacefully farms fungi in healthy eucalyptus trees in southwestern Australia. "They have a totally controlled environment and are very protective of trees," Dunn says. The ambrosia beetles, led by a queen, excavate tunnels in the heartwood, where they garden fungi and lichens for decades. The daughters of the founding queen, a group of infertile Amazons, clean and defend the garden beetle society with the devotion of cloistered medieval nuns. They never leave the tunnel. "The real value of looking at bark beetles," explains Dunn, "is that they are a fascinating form of life. They can also be the most destructive form of insect life for sheer scale of effect on the planet. And they can change a forest unlike any other creature. That makes them an important species to pay attention to."

Together with the physicist James Crutchfield, a specialist in chaos theory, Dunn worries that bark beetles may also become novel players in global warming. Crutchfield owns some land in New Mexico, where beetles took out nearly a hundred trees, and has partnered with Dunn on a number of music and beetle projects. The chaos theorist and the musician calculate that the bark beetle has the potential to be not just a quick responder to higher temperatures but a generator of carbon, based on a number of feedback mechanisms. Their reasoning goes like this: As more trees die, less carbon is sequestered and stored. The disappearance of forests, in turn, leads to the generation of less oxygen and the concentration of more carbon in the atmosphere. As the release of more forest carbon causes temperatures to rise further, the warming will put beetles on the landscape in ever-greater numbers and ever-expanding geographies. Crutchfield and Dunn call it entomogenic change. "Things have the potential to run away on us." says Dunn. "Things that seem in balance may go off the scale."

It is not an idle theory. The drought and the beetle die-off in the pinyon juniper forest changed the region's entire carbon budget. According to one 2010 study, the U.S. Southwest lost more carbon from the beetle/drought die-off (an estimated 5 million tons) than it did from wildfire or logging over the same period. Spruce, lodgepole, and Douglas-fir forests generally hold two to three times as much carbon as the slow-growing pinyon. But beetles have dramatically whittled the capacity of those trees to hold carbon.

Dunn doesn't know where the bark beetle will take his own art or how the insect will ultimately redesign the earth's forests and climate. He hopes that sonic warfare will eventually calm the beast in the forest. But the artist knows that he has surely met a charismatic creature like no other, a Beethoven among insects. If nothing else, the beetle should remind us, says Dunn, that we are all part of nature; we come from the natural world and will live or die depending on our generous understanding of that world. Like the Babylonian beetle empires now rising and falling in our forests, human civilization has abruptly experienced collapse, and will again. The problem with nonlinear change (and neither human nor bark beetle empires move in a straight line) is that "there is no off switch," says Dunn. "We can't reboot the software." He suspects that the sound of the beetle in drought-wounded trees heralds profound transformation. He pauses before he adds a final thought. "The changes probably won't be good for us."

EIGHT

The Sheath-Winged Cosmos

. . . . .

"Dar'st thou die?
The sense of death is most in apprehension,
And the poor beetle, that we tread upon,
In corporal sufferance finds a pang as great
As when a giant dies."

WILLIAM SHAKESPEARE, *Measure for Measure*

BEFORE THE advent of the computer screen, suicide bombers, and acid-filled oceans, beetles got some respect in this world. When human languages and cultures were as diverse as the stars, beetles informed how we lived. They inspired artists, fascinated scientists, illustrated evolution, educated philosophers, pollinated crops, delighted children, hardened warriors, decorated women, and put food on the table. They even merited legal representation in fifteenth-century ecclesiastical courts. Not so long ago, the noble beetle populated our songs, dreams, poems, proverbs, and fables. European peasants gave thanks to "Beasts of the Virgin." Egyptian pyramid builders wore dung-beetle charms for good luck. Samurai warriors strutted about like Japanese horned beetles, and German and French villagers made soups out of cockchafers. Beetles inspired all sorts of human inventions, including agriculture, the wheel, mummification, the theory of evolution, and, yes, the chain saw.

Although most politicians don't know it, beetles belong to the "great commonwealth of living things." Coleoptera (the word means "sheath-winged") not only outnumber and outrank mammals but predate dinosaurs. They remain the most successful animals on earth. When William Blake wrote about holding "infinity in the palm of your hand and eternity in an hour," he was probably thinking about beetles. Every day, the sheer audacity and abundance of beetles makes *Homo sapiens* look like a somewhat spindly branch on the tree of life.

Beetleness is about being small, six-legged, invisible, mobile, and well protected. Unlike such insect show-offs as butterflies, beetles generally live out of the way, beneath the surface of things. Every beetle begins life as an egg and then progresses to a wormlike state (larvae and pupae) before metamorphosing into an armored adult sporting a fancy pair of forewings with protective hard casings called elytra. Many beetles fly as clumsily as drunken sailors walk.

In the global food chain, beetles remain our greatest and sharpest competitor for nutrition. Grain weevils (members of a family of numerous species of beetle with a comical long snout) typically chew their way through a third of the world's grain crop every year. The Western corn rootworm, dubbed "the billion-dollar beetle," can reduce corn yields between 10 and 80 percent. Cucumber beetles, potato beetles, taro beetles, and soybean beetles all nibble at the global food banquet. Beetles are hardcore foodies. "For every bean full of weevils God supplies a blind grocer," goes an Arab proverb.

Since 1758, scientists have described and named more than 400,000 species of Coleoptera. That's more than four species a day. In 2005, U.S. scientists named two new species of beetles that dine on slime mold, *Agathidium bushi* and *Agathidium cheneyi*, after then president George W. Bush and his VP Dick Cheney. (There's an *Agathidium rumsfeldi* too!) When the entomologist Terry Erwin fogged the canopy of just one tree species in a Panama jungle in the 1980s, 163 different species of beetles dropped to the ground. Coleopteran exuberance once prompted the irreverent British geneticist J.B.S. Haldane to suggest to a group of clerics that God obviously had "an inordinate fondness for beetles."

God may even look like a beetle, for in diversity lies not only divinity but extraordinary resilience. Coleoptera make up a third of all life on the planet, and a quarter of all animals. If the Internet accurately reflected biological life on earth, one out of three websites would be devoted to beetles, including pugnacious beetle sex. Most scientists suspect the actual number of beetle species is close to 10 million. Not surprisingly, beetles live in rivers, lakes, jungles, caves, forests, and deserts and on mountaintops. Bark beetles alone outnumber all mammal tribes by at least a thousand species. In fact, the great Coleoptera fraternity makes mammals look like a quaint evolutionary experiment with limited prospects. Thanks to climate change, forest destruction, and human city making, a quarter of all mammals may go extinct by 2050. The beetle tribes that survive such depredation may well inherit whatever is left.

Coleoptera stand out as walking evolutionary billboards for innovation and adaptability. Scientists once thought that beetles' numerical superiority sprang from their size, their mobility, and their well-protected bodies. But their impressive diversity, still the subject of rigorous debate, probably owes as much to their evolutionary age as to their collective ability to change their diet.

Beetles first appeared in the fossil record 350 million years ago, and over time evolved into more than one hundred separate families that largely fed on primitive grasses, mosses, palms, fungi, and the bark and wood of gymnosperms. These forebears to modern trees included pines, conifers, and ginkgoes. Some bark beetles still dine on the monkey puzzle tree in Argentina right next to fossils of their 200-million-year-old ancestors. Scientists call this beetle picnic area "a little Triassic Park."

When flowering plants appeared some 140 million years ago, beetles took advantage of this new fast food and exploded in their diversity. Some beetle species dined on flower petals; others chewed leaves, stems, fruits, and roots. (A beetle's ability to turn a plant or tree into a leafless or fruitless skeleton is really an insect version of shock and awe.) More plants begat more beetles, and more beetles begat more plants. Wherever you find lots of beautiful flowers—Papua New Guinea and Venezuela are two examples—you find an abundance of beetle species. (And a diversity of human communities, too.)

Beetles and plants, of course, have also been at war for millions of years. Every time a plant secreted a new poisonous sap or perfume to ward off hungry beetle armies, the insect adapted by making its own defensive chemicals derived from its herbaceous enemies. Although half of all beetle species remain dedicated plant eaters, many have changed their eating habits to include excrement (digested plant matter), insects, fellow beetle species, vertebrates, and carrion. Ground beetles and rove beetles, for example, behave like the jaguars of the insect world and eat whatever comes across their path. Dermestids specialize in cleaning up bones, so natural-history museums keep colonies of the flesh eaters on hand to polish up animals with backbones.

Scavengers such as the famous dung beetles probably started off dining on the excrement of mammal-like reptiles before moving on to mountains of dinosaur shit. Today, their well-diversified descendants (some 7,000 species) work with camel, elephant, kangaroo, or howler monkey dung. In just two hours, 16,000 dung beetles can clean up and bury 3.3 pounds of elephant poop.

Some members of the dung beetle family gather at the anus of a marsupial in anticipation of an offering, much like Sunday urbanites lining outside a breakfast diner. Other dung beetle species dine on the slime tracks of snails.

Beetles regulate the common wealth of trees and other plants by safeguarding diversity. Their innumerate duties include gardening, dissembling, pollinating, boring, pruning, killing, recycling, and refuse eating. They are Mother Nature's handiest garbage collectors, and as such belong to the prestigious FBI agency of global dissolution: fungi, bacteria, and insects. As engineers of decomposition and global protein renewal, beetles take apart weak, abundant, or aging plants, thus making room for new growth. They break down the detritus of ordinary life: windblown spruce, dead birds, fallen leaves, rotten apples, and every kind of foul emanation. Insect researchers, who sometimes sound like car mechanics, describe beetles as agents "crucial for ecosystem function." In plain English, the world would be an undeniably odious place without them. Without beetle omnivores, most landscapes would be full of dead things.

As both the spruce and the pine beetle ably illustrate, bark beetles occupy a high rank in the great commonwealth: that of chief forest manager and tree surgeon. Their specific duties include decomposing dead trees, silencing damaged or diseased trees, pruning the injured, and killing aging forests on the verge of collapse. A bark beetle is to an old tree what pneumonia used to be to an old man: his best friend. By taking down entire forests, bark beetles and their associates typically restore diversity, recycle nutrients, and generally shake up the established order of things. To armies of bark beetles, an aging forest looks like a decadent empire bereft of energy. No animal on earth other than humans can change a landscape as dramatically or as quickly as bark beetles.

But Coleoptera graduates all sorts of environmental engineers, too. Consider for a moment the celebrated dung beetle of the scarab family. These hard-working animals aerate the soil, sow undigested seeds, recycle nutrients, and suppress other bothersome insects. These beetles live on dung, mate on dung, and lay eggs on dung. They see every pile of waste as a gift. Every year, dung beetles perform nearly $60 billion worth of excrement removal on North American livestock pastures alone. Without scarabs, our soils would be infertile and our waters would be choked with algae and fecal matter.

Australia learned the hard way about the value of the dung beetle. When Europeans transformed that continent into an imperial cattle kingdom by importing the four-legged alien more than two centuries ago, cow dung accumulated in unsightly piles. Native dung beetles could handle dry kangaroo pellets but couldn't deal with wet cow pads. As a result, the cow dung (half a million tons a day) polluted water and became fertile breeding ground for native bush flies. One pad alone could incubate about two hundred irritating flies. Black clouds of pesky bush flies plagued citizens in the outback: "We can't help eating, drinking, and breathing flies; they go down our throats in spite of our teeth and we wear them all over our bodies." In 1967, the Australian government introduced more than fifty species of dung beetles from Africa and South America. These cow-pie specialists quickly deprived the Australian bush fly of breeding piles and enriched the soil. Bush fly populations dropped from an incredible eight thousand bugs per acre to four hundred.

Beetles are cold blooded, and thus owe their evolutionary success to a high level of climate literacy. Together with most insects, beetles respond promptly to cold or warming spells. Extinction doesn't appeal to Coleopterans. Thanks to their mobility (most can fly), brief life spans, and wild reproduction rates, beetles can relocate swiftly. They can expand or contract their neighborhood as well as their breeding time. For these reasons, beetles provide some of our best fossil evidence on how insects respond to changes in climate. The remains of hard-wing beetle casings show that the cold-loving beetles that live in the Arctic today occupied Liverpool, England, nearly 15,000 years ago. When England was much warmer, about 120,000 years ago, dung beetles dined on shit piles left by herds of roaming bison. These same species now reside only in southern Europe. Faced with global cooling or warming, beetles, and in particular bark beetles, are well equipped to either expand their range or pack their bags. They are also sentinels of climate change.

Beetles pioneered agriculture as well, unbelievable as that may sound. Only four insect academies in the world have mastered farming, and bark beetles rank first among them. About 60 million years ago, ants learned how to cultivate gardens of fungi in the Amazon rainforest. Termites became fungal peasants in the African jungle about 35 million years later. But ambrosia beetles (there are 3,400 species in the subfamily Scolytinae) started fungal farms in trees not once but seven different times during the last 60 million years. (Humans didn't starting planting grains until 10,000 years ago, and we still work like hell to keep fungi out of our crops.) Early insect watchers named these beetles after the food of the gods, ambrosia, because they couldn't figure out what they ate.

Ant, termite, and beetle farmers could choose from a variety of fungi to cultivate, because mushrooms don't require light or pollination. Among ambrosia beetles, only the females farm. They tunnel into bark or wood, carve out small plots, and set up tidy gardens consisting of fungi, yeast, and bacteria. Bacteria controls diseases on the fungal crops. When the beetle brood hatches, they feed on the nutritious garden. Female beetles inspect, prune, and monitor their gardens with the infectious intensity of a Gertrude Jekyll or Vita Sackville-West. The 50-million-year success of beetle agriculture suggests that human farming, especially given our crude industrial practices, hasn't found its proper scale, place, or tools.

The prowess of beetles invites awe in numerous ways. Some species have broken more world records than any other animal. The Australian tiger beetle, for example, the world's fastest insect, chases down its prey at .53 meters a second. (That speed is almost nine times faster than an Olympic sprinter in terms of relative body length.) The African bombardier beetle squirts a jet of boiling liquid (212 degrees Fahrenheit) containing hydrogen peroxide at its enemies at 300 pulses per second. The energy mustered by a firefly to create light (three families of beetles are known for their bioluminescence) is nearly a hundred times more efficient than an incandescent lightbulb, all in the name of good beetle sex. The rhinoceros beetle, the world's strongest animal, can effortlessly ferry 30 times its own weight and carry up to 100 times its weight (one-tenth of an ounce) with difficulty. Science writer Carl Zimmer compares the latter feat to "a 150-pound man walking a mile with a Cadillac on his head without tiring."

Whereas the feather-winged beetle, among the world's tiniest insects, is small enough to crawl through the eye of a needle, the Goliath beetle can fill the palm of a human hand. In the forests of Papua New Guinea, one inventive weevil genus with ten or so species carries a fungus garden on its back, which in turn houses roundworms, mites, and bark lice. Males of the death-watch beetle can eat their way through old oak timbers for ten years, making a clock-like tapping sound to attract mates. The desert darkling beetle exudes a special kind of wax to keep itself from drying out in the hot sun. To move a dung ball in a straight line, a dung beetle senses polarized light from the moon and the sun, something no human can see. (On overcast days, dung beetles travel in circles.) Some male beetles ejaculate up to 10 percent of their body weight and are a source of water for females. The infrared sensors on a black fire beetle can help it locate a forest fire several miles away. (On smoldering branches, the beetles copulate without fear of predators and lay their eggs in warm wood.)

The shell of the Asian Cyphochilus beetle possesses a whiteness brighter than milk or the human tooth. The endangered American burying beetle, which once dined on dead passenger pigeons, can sniff a mouse or quail carcass from two miles away within an hour of death. The beetle buries the dead animal and then supports its family on the carcass preserved by an antibacterial secretion that slows rotting. Larder beetles can reduce a dead human in an apartment to a skeleton within five months and are much prized by forensic scientists. After Joseph Cox spotted the sharp, C-shaped teeth on the larvae of a wood-boring beetle (*Ergates spiculatus*), he started the Oregon chain-saw company and made history. The marvels of beetles are as endless as the stars, and many are still beyond our comprehension.

Until recently, almost all peoples on the planet ate beetles because of their abundance and high protein value (more than that of beef or pork). Native gatherers and poor farmers routinely included beetles on the menu. The Betsileo of Madagascar ate cockchafer grubs. The Bedouin roasted scarabs, and women of North Africa gobbled up beetles to make them more plump. Until the nineteenth century, the finest French restaurants served chafer bouillon. Thais soaked beetle larvae in coconut curries and then roasted them. South Americans spiced up their diets with larvae of palm weevils and passalid beetles. A Russian once said, "In the steppe even a beetle is meat."

Beetles remain delicacies in various parts of the world, particularly Africa and Asia. The Chinese prepare water beetles by pulling off the elytra and the legs, then fry the insects and eat them like nuts. The Vietnamese dip spring rolls in a sauce that contains the elixir of a rare water beetle called *ca cuong*. In the South Pacific, the grub of the palm weevil, traditionally battered and fried or roasted over a grill, is thought to taste like beef marrow. Collected from the base of a banana tree, the large larvae of the Goliath beetle make a fast-food treat in much of Africa.

ALMOST EVERY human civilization except our own has recognized beetles as charismatic creatures. When ancient Egyptians watched scarabs roll their balls of dung across the sand, they saw their own religious world re-enacted in miniature. To them, the dung beetle's work looked a lot like that of Khepera, the creator and god of the rising sun, daily pushing the sun across the sky. The Egyptians not only embalmed scarabs along with mummies but believed that these sacred bugs embodied the essence of *cheper*: the idea of becoming, existing, and revolving. Indeed, the practice of mummification probably arose from people watching the hooded-like pupae emerge from their pupal cases. Egyptians from all walks of life wore the likeness of scarabs on amulets, jewelry, and seals for good luck and long life. Indian tribes in Bolivia's Gran Chaco country still believe that a scarab named Aksak created man and woman out of clay.

The German biologist Gerhard Scholtz suspects that dung beetles inspired humans in the Middle East to invent the wheel. As rural folks domesticated more cows, goats, and sheep, the activity of scarabs grew right along with the dung piles: "It was almost impossible for their owners to overlook the curious activities of dung beetles." A person pushing a wheeled cart still looks much like a dung beetle rolling a ball of dung.

To the Greeks, beetles symbolized "everything from industriousness to insignificance." By watching fireflies, Aristotle learned that "some things produce light" even though they are not on fire. Peasants in particular cherished the popular Aesop's fable about the eagle and the dung beetle. It goes like this: When a hare sought asylum with a dung beetle, an arrogant eagle brushed the beetle aside and ripped the hare apart. Outraged, the beetle flew up to the eagle's nest and twice destroyed the bird's eggs. The distraught eagle sought to protect his next batch by leaving the eggs in the lap of mighty Zeus himself, on Mount Olympus. The beetle was undaunted. After rolling in ripe excrement, it flew right into Zeus's face. The rattled god leapt up, spilling and breaking the eagle's eggs. Every Greek peasant got the moral: the vilest beetle can outsmart the mighty.

During the Middle Ages, amazingly, church authorities in France and Switzerland often put beetles on trial. Crop eaters, tree defoliators, and other swarming insects routinely became the subject of celebrated ecclesiastical cases administered by conjurers and argued by famous jurists. The Church entertained two views on insects: it either condemned the bugs as emissaries of the Devil or saw them as servants of a wrathful deity that demanded penance and prayer.

Judging a beetle in court was a curious business that required elaborate protocol. After a cloud of insects had afflicted local crops, peasants could petition authorities for redress. In turn, clerics would appoint lawyers to represent the swarming insects. After the insect lawyers had filed their pleas, lawyers for the starving peasants could respond. Eventually a cleric would pronounce a sentence on the bugs. Such insect trials were "more heavily weighted with the spoils of erudition than the set speech of a member of the British Parliament," writes E.P. Evans in his extraordinary book, *The Criminal Prosecution and Capital Punishment of Animals*. Clerics often showed more wisdom than modern forest officials do in their judgments: "We cannot tell why God has sent these animals to devour the fruits of the earth; this is for us a sealed book."

Insect defendants invariably included locusts, weevils, and cockchafers, a member of the scarab family. The voracious cockchafer, now subdued by pesticides and expanding cities, once dined on tree leaves and the roots of valuable wheat crops. Chafers assembled in such large numbers that their drowned bodies clogged river watermills. They commonly appeared on French soup menus and populated German nursery rhymes. Chafers would bang noisily on windows, drawn by the flames of candles.

After an infestation unsettled the good citizens of Berne, Switzerland, in 1478, a magistrate ordered the bugs to "appear before the bishop in order to tell their story." Although the chafers failed to show up, the bishop later swore that he'd heard their "vicious and abominable answer."

In 1545, "a greenish weevil" infested the vineyards of Saint-Julien, a small French village near Mont Cenis. When the winemakers complained loudly, the beetles were assigned a seasoned lawyer. Two years later, an official called for public prayers instead of rash action against the accused, on the grounds that "the supreme author of all that exists hath ordained that the earth should bring forth fruits and herbs not solely for the sustenance of rational human beings but likewise for the preservation and support of insects, which fly about on the surface of the soil."

After several high masses and much public penitence, the beetles abandoned the fields. But thirty years later, the weevil returned, and this time the accused had their day in court, a process that took more than forty years. The winemakers appealed to the Church to stop the beetles' "inordinate fury" and "incalculable injury" with prompt insect excommunication. The appointed beetle advocate replied that the accused were only "exercising a legitimate right conferred upon them at the time of creation" and that civil law did not apply to "brute beasts," because they followed only natural law. Not surprisingly, the advocate argued that the plaintiffs' petition be dismissed. Lawyers for the grape growers replied that animals had been created for man and were supposed to be "subordinate to him and subservient to his use." Given the extraordinary length of the legal proceedings, the local commune graciously offered to provide a well-guarded asylum for the weevils outside the vineyard. Defense counsel for the beetles rejected the location as "sterile nor suitably supplied with food." The outcome of the case remains unknown today, because rats or bugs chewed up the final page of the judgment. As

E.P. Evans sarcastically concludes, "Perhaps the prosecuted weevils, not being satisfied with the results of the trial, sent a sharp-toothed delegation in to the archives to obliterate and annul the judgment of the court."

The medieval mind, however, could accommodate beetles in a way no modern economist can. Remarkably, all parties to the case described believed that insects had the right "to adequate means of subsistence suited to their nature." In an 1846 treatise on these unique insect trials, French author Leon Menabrea concluded that they represented something strangely noble: "In the Middle Ages when disorder reigned supreme, when the weak remained without support and without redress against the strong and property was exposed to all sorts of attacks and all forms of ravage and rapine there was something indescribably beautiful in the thought of assimilating the insect of the field to the masterpiece of creation and putting them on an equality before the law. If man should be taught to respect the home of the worm, how much more ought he to regard that of his fellow man and learn to rule in equity."

During the time beetles still had recourse to courts and counsel, the ladybug became a legendary figure throughout Europe. When a cloud of the colorful beetles answered fervid prayers to the Virgin Mary to attack crop-destroying aphids, farmers thereafter honored the bug as "the Beetles of Our Lady" or "Lady Beetles." In France they become known as "Cows of the Lord" or "Beasts of the Virgin Mary." Norwegian farmers called them "Our Lady's Key-maid."

When a ladybug landed on a young woman's hand, she thought the beetle was measuring her for wedding gloves. In Scotland, a ladybug could divine "one's future helpmate"; the spots on its back even indicated how many children you might have. In the Netherlands, the beetle promised good luck. When British peasants burned hop vines after the harvest to clear the fields, they sang, "Lady-bird, lady-bird, fly away home. Your house is on fire and your children all burn." Doctors treated measles with lady-bugs and rubbed the beetle on cavities to stop toothaches.

The medicinal properties of beetles represent another area of lost traditional knowledge. Before megacorporations existed, ordinary people relied on beetles in what is now a disappearing pharmacopia. During the eighteenth century, men purchased a love potion made from the secretions of the blister beetle (cantharidin) to improve their sex lives. (An overdose created a permanent erection or death.) The blister beetle also releases a chemical still used in a popular wart removal. A stag beetle tied around the neck prevented bedwetting. Surgeons used the mandibles of certain beetles to suture wounds. The antibacterial properties of dung beetles cured earaches. The Chinese used to pop ladybugs, which contain natural painkillers, to ease their proverbial aches. Malaysian folk remedies for back pain and respiratory disorders still call for the ingestion of fifteen live beetles (which dine on herbs) in a glass of cold water. A growing number of Argentines have started to swallow Asian darkling beetles to treat cancer, AIDS, and diabetes.

Beetles have made history in other fantastic ways. Two thousand years ago, Aztec and Inca farmers harvested female cochineal beetles, full of carminic acid, from the prickly pear cactus to make a highly prized and luxurious red dye. Armed with sticks and brooms, the Indians would painstakingly remove the beetles from the cactus and then boil them or bake the insects in the sun for several days. To make one pound of dye required the drying of seventy thousand beetles. Every year in total, villages sent a tribute of ten tons of cochineal to the Aztec emperor, more than 1 billion beetles. The Aztecs so cherished the color red that they dyed feathers, painted buildings, and illustrated books with cochineal. Women strikingly decorated their hands, neck, and breasts with beetle juice, and prostitutes reddened their teeth with the dye. The ancient farmers of Mixteca and Oaxaca tended to cochineal beetles with such fastidious care that they created a domestic variety twice the size of the tiny bug's wild cousins.

The richness and durability of the cochineal beetle dye astonished Spanish conquistadores as well as the European fashion trade. Old World elites also fancied red as a power color, but no existing European dye could match the vividness of cochineal. The Spanish expropriated the Aztec industry and turned it into an empire money maker second only to silver exports, keeping the dye's insect origins largely a trade secret until the 1900s. By the seventeenth century, cochineal appeared on stock-market listings, and merchant bankers ruthlessly speculated in the trade. White-faced queens shocked their vassals with cochineal-stained lips, and bakers added ground beetle bits to "outblush the cherry and the plum." European doctors hailed the bug as a fine antidepressant, good against "melancholy diseases, vaine imaginations, sighings and griefe." Mental patients dined on meals reddened by the beetle to calm their nerves. To paint "a Colour that always looks most Noble," Rembrandt, Vermeer, and Turner relied on cochineal. For several hundred years, every king, pope, and idol of the Virgin Mary dressed in resplendent red from *grana cochinilla*. During the eighteenth century, cochineal mixed with tin made a "perfect Scarlet" that soon colored Scottish kilts as well as the British Empire's "redcoats." Though cancerous synthetic dyes eventually eroded the cochineal trade, the beetle dye is still used to color lipstick, candy, fruit juice, applesauce, sausages, and Campari. Just about everyone has tasted cochineal. The famous ecologist James Lovelock once combined a 112-pound sack of the beetles with acetic acid and copper to produce a red so intense that "it seemed to draw the sense of color out through my eyes from my brain."

DURING THE nineteenth century, scientists and explorers talked about beetles the way people today might enthuse about cars or sport teams. The Victorians regarded beetles as a window on the mystery of creation and as distraction from the ugliness of industrial society. The great naturalist Henry Bates wrote about them, as did Baron Alexander von Humboldt. Humboldt, an enterprising explorer and naturalist, once watched Cubans fill a translucent gourd full of fireflies to make light for the evening. The poet of entomology, Jean-Henri Fabre, wrote an entire book about weevils. He also found the cooperative behavior of dung beetles morally instructive: "Family burdens that would exceed the strength of one are not too heavy when there are two to bear them."

Both Alfred Wallace and Charles Darwin, co-discoverers of evolution, came up with the radical idea of natural selection based, in part, on detailed observations of beetles and beetle sex. Both scientists collected the insects as boys, and Darwin's zeal for beetles seemed irrepressible: "One day, on tearing off some old bark, I saw two rare beetles, and seized one in each hand; then I saw a third and new kind, which I could not bear to lose, so that I popped the one which I held in my right hand into my mouth. Alas! it ejected some intensely acrid fluid, which burnt my tongue so that I was forced to spit the beetle out, which was lost, as was the third one."

Wallace and Darwin both traveled to the tropics, where the diversity of beetles and other creatures essentially taught them that natural selection drove evolution along. Animals (including human tribes) adapted to their neighborhoods over time by selecting the most successful color, size, or habit for survival. Wallace, one of the first scientists to link carbon dioxide to climate warming, also maintained that if scientists didn't collect, study, and name tropical beetles, then "future ages will certainly look back upon us as a people so immersed in the pursuit of wealth as to be blind to higher considerations." Darwin deemed the Atlas beetle, "with its polished bronze coat of mail and its vast complex of horns," the world's most impressive animal.

During nineteenth-century beetle mania, the insect became

a popular jewel and fashion statement. Fashionistas couldn't get enough of beetles, butterflies, or hummingbird feathers. The wings of jewel beetles even decorated tea cozies. Victorian women bought showy fans made of peacock feathers and beetle parts as well as lace dresses composed entirely of beetle wings, stitched in elaborate leaf and flower motifs. At a Bengal bazaar two hundred years ago, an eighty-two-pound bag of shining green buprestid wings fetched six rupees for future use in textile decoration. Wealthy ladies from India, Mexico, and Sri Lanka kept inch-long iridescent beetles as living jewels to be worn on important occasions. They bathed and fed their precious armored ornaments, which they kept in cages. The headhunters of western Amazonia traditionally decorated themselves with human hair, toucan feathers, and copper-coloured beetle wings, or elytra. The Karens, a Thai hill tribe people, attended funerals wearing "singing shawls" decorated with long, rattling beetle elytra.

Beetles have also given the business class some pointed lessons. After cotton was introduced to the United States in 1786, the plant soon colonized the South, with grand plantations worked by black slaves. Bankers dressed in Sea Island cotton celebrated the predictability and efficiency of a cotton economy. Then along came the boll weevil, a vegetarian that chewed on cotton blossoms and could reproduce ten generations in one year. The insect eventually did what the Civil War failed to do: mortally wound the U.S. cotton economy.

A native of Mexico, the boll weevil first chewed its way across the Rio Grande in 1892 and was soon attacking plantations throughout Texas. Experts thought the bug wouldn't cross the Mississippi River, but it did. Within 30 years, it had spread over 600,000 square miles of cotton fields. By 1910, the blues singer Charley Patton was singing the "Mississippi Bo Weavil Blues": "Bo weevil, bo weevil, where your native home? Lordie. 'Most anywhere they raise cotton and corn.' Lordie."

Before the insect arrived in Alabama around 1914, that state had cropped a record 1.7 million bales of cotton. In short order, a beetle infestation slimmed the harvest by 70 percent, putting bankers, plantation owners, and sharecroppers out of business and sending black cotton pickers north to industrial cities. Scientists tried drowning weevils in kerosene and carbolic acid, but the weevil kept on coming. (Delta Airlines got its start in the aerial business by spraying arsenic dust on weevils.)

To stem the beetle invasion, despairing farmers in Coffee County planted only peanuts, corn, and potatoes. Crop diversity revitalized not only local soils but the entire farm community. Across the south, the insect freed the economy from the stranglehold of King Cotton. The grateful citizens of Enterprise, Alabama (now nicknamed Weevil City), erected an Italian-made sculpture first depicting a lady holding a fountain. The sculpture was later replaced with a weevil. The monument read "In appreciation of the Boll Weevil and what it has done to the Herald of Prosperity." Although repeatedly vandalized, it remains a testament to the power of the ancient order of Coleoptera to change human thinking.

It is doubtful any logging community in the U.S. or Canada will erect a statue to the pine or spruce beetle. Yet bark beetles have unsettled corporate logging communities more assuredly than the boll weevil undid the cotton economy. No North American court would give beetles the time of day, yet scolytids have an evolutionary rule to perform in the forest. People once appreciated the diversity of beetles and lived alongside them as fellow members of creation. Human societies embraced their presence. The less we know and appreciate about the natural world, the narrower and darker our lives will become.

That's how George Ball sees things, and he's been studying beetles all his life. "They appeal to the simple mind," says the eighty-four-year-old Canadian ground beetle expert. "It's the aesthetics. They have a nice compact form. You get such a variation of shape, size, and color, too." Ground beetles caught Ball's interest as a young man, and the family's forty thousand known members still dazzle him. "I like 'em all," says the Detroit-born Ball. The renowned coleopterist wears a belt with a bronze tiger beetle buckle and a shirt embroidered with a bombardier beetle. Beetle specialists tend to be parochial animals, he explains.

Carabidae, of course, are a wonderful family. In addition to burning his fingers with hydrogen peroxide (the chemical cocktails some bombardier beetles squirt out of their rear ends are strong enough to knock over a small mammal or cause heart failure), ground beetles opened the world up for Ball. "They taught me about life's great diversity and gave me a sense of wonder. They are a grand illustration of natural selection. Without the beetle, we would have a poor idea what the possibilities are for adaptation."

The Carabidae, the seventh largest family of beetles, perform specific duties that speak of ancient realms and unmapped traditions. "They're mainly predators that clean up other insects," says Ball. But most of the carefully pinned samples in Ball's tidy drawers at the Strickland Entomological Museum are from species occupying niches no human truly understands. As the emeritus curator carefully opens a drawer of luminescent *Calleida* and another containing the ant-like genus *Lebia*, he confesses, "We don't know much about them," though many Carabidae move, as Ball puts it, as "fast as streaked lightning." The jet-black *Pasimachus* may or may not eat earthworms. "It's always a good day when you get a *Pasimachus*," says Ball. "They are not that easy to find."

Ball calls bark beetles a "wonderful group" but adds, "Bark borers look all the same to me." Unlike ground beetles, which live simple lives ("it's hunt and grab"), bark beetles are quite highly evolved and extremely social. "They can communicate by sound and odor. They have the makings of a society."

Although thousands of new ground beetles are discovered in disappearing forests, deserts, and grasslands every year, "there are not enough people to identify them." Most of the new beetles go unnamed. Nor are there enough scientists being trained "to tell us what the hell it is that is being lost," says Ball.

In the great scheme of things, Ball sees humans as just one of many species. "We should care about beetles just as we should care about our environment," he says. "When we impoverish the environment and make it no longer self-sustaining, we are not really looking after beetles. But they're looking after the environment we depend on." Every time we eliminate another beetle species from the world, adds Ball, we are in a small and incremental way "exterminating ourselves." The ground beetles that inhabit the forest or live by standing water form an interconnected web. "If you start to knock out pieces here and there, you'll sooner or later cause a major upset and start things down the slope of extinction."

Ball views beetles as fitting reminders that small is beautiful. And the prism of colors on an animal's elytra wings can be "heart-stopping." Beetles have given him "just a marvelous way of making a living." They are also a living testament to the astonishing mystery of this world. The geneticist J.B.S. Haldane, who relished God's fondness for stars and beetles, wrote that "the universe is not only queerer than we suppose, but queerer than we can suppose." Coleoptera, the greatest branch on the Tree of Life, proves that fact every day.

NINE

The Two Dianas

· · · · ·

"It is not so much for its beauty that the forest makes a claim upon men's hearts, as for that subtle something, that quality of air, that emanation from old trees, that so wonderfully changes and renews a weary spirit."

ROBERT LOUIS STEVENSON, "Forest Notes"

IN APRIL 2010, Diana Six, the beetle expert her colleagues call "the fungal guru," grabs a hatchet and whacks off the bark on a felled lodgepole at the Lubrecht Experimental Forest. Located just thirty miles northeast of Missoula, Montana, the study area is the largest outdoor tree classroom and laboratory in the United States. But *Dendroctonus ponderosae* has rearranged the place just as it redesigned Six's front yard, from which the scientist had to remove eighty ponderosa pines. As in much of Montana, piles of beetled logs litter the dry experimental forest floor like pick-up sticks. In an attempt to stop the unstoppable, authorities decorated the entrance to Lubrecht with a massive clear-cut. But the beetles just kept on coming.

Underneath the bark of one felled lodgepole, Six quickly finds the evidence she's looking for: various stages of beetles and pupae in J-shaped galleries. She points out mature darkened beetle adults, all ready to fly a month ahead of schedule, then performs the crunch test: "If they squirt, they're alive." The beetles pass with flying colors.

"Damn, this is early," she says energetically. "It's April. This is the earliest I've ever seen. But I'm shock-proofed now." She suspects some of the beetles have already left in search of new wooden estates. In addition to the young adults, she's also spotted year-old veterans ready to go it again. "Normally they would mate and then die over the winter," she explains. "Now, quite a few make it through and emerge to fly off and kill more trees." It's not unusual for female beetles to wake up in the spring and lay yet another batch of eggs. "These guys are ready to pop. Amazing."

The clock that once determined the size of a beetle empire has been broken, Six says. "Our guidelines on beetle flights were based on temperature, elevation, and food supply, and they're gone. We have beetles completing various stages of development at different times and flying out at different times. The beetles now go to the tops of mountains."

But Six suspects that climate change could conceivably unravel the very insect empire it propelled into prominence. The pine beetle, for example, depends on two different fungal species to provide essential food in a tree. *Grosmannia clavigera*, an ancient passenger on the beetle bus, provides a high-quality superfood, while *Ophiostoma montium*, a newer addition on the evolutionary timescale, supplies an adequate dinner. Larvae gorge on the nitrogen-packed strands of fungus that grow like spiderwebs into the phloem and wood, while adults dine on the sterol-packed spores. Without a parting fungal meal, young beetles can't really attack trees, carve galleries, or lay eggs. "Fungi are what make the mountain pine beetle the mountain pine beetle. We wouldn't have this beetle explosion without the fungi. It just wouldn't be around or able to do what it does."

Six, along with researcher Barbara Bentz, recently discovered that *Grosmannia* and *Ophiostoma* thrive in different temperature ranges. The beetle ensures its dietary needs by cleverly switching from one fungal species to another as the thermostat goes up or down. When temperatures range around 77 degrees Fahrenheit or below, the beetle transports the hardy spores of *Grosmannia clavigera*. When things warm up to between 77 and 100 degrees, the beetle drops the G. fungus like a hot potato to exploit the dark-colored *Ophiostoma montium*. Beetles that fly early in the summer tend to carry more *Grosmannia* than *Ophiostoma*, and late-summer flyers might exclusively transport the O. food package, says Six. But if global warming continues to heat up the West, it's possible that *Ophiostoma montium* might not be able to take the heat. The fungi have the same generation length as the bark beetle, as well as their own limits on adaptability. "We know that *Grosmannia* will drop out first, and next comes *Ophiostoma*. Then the beetles will be in real trouble unless they can find a new fungus."

Such decoupling is not a new phenomenon. Global warming, combined with industrial pollution, has a way of severing odd and opportunistic marriages in the natural world. Coral reefs used to have a steady relationship with a particular alga that helped "the tropical rainforests of the ocean" gather food and energy. The algae also gave the coral complexes their resplendent colors. But as the oceans got more polluted, acidified, and warmer, the algae died or bailed out. Left high and dry, the coral bleached. A decoupling, Six explains, "is not a gradual development or linear response. It hits a threshold, and things go over the edge." If the bark beetle can't find another mobile fungus or an existing companion that can genetically adapt to hotter summers, the pine beetle could lose its moxie in some forests. "We are trying to figure out when and where it could happen." A pine forest without a beetle would be like a whitebark without a Clark's nutcracker: it would be a poorer place.

Every year, Six travels to South Africa to study fungi and beetles. There too, the trees seem to be departing in hellish numbers. The poison arrow tree, a giant cactus-like creature (*Euphorbia ingens*), used to decorate sandy, rocky-topped hills and provide food for black rhinos and baboons. But now hillside after hillside is turning gray as "the trees mush out." In some areas, 90 percent of the trees are dead or dying. Six and other researchers looked for a pattern but couldn't find one. "Yes, the tree was getting hammered by beetles and fungi, but they were just responding to a distressed tree like ambulance chasers," Six says. She suspects that global warming has shifted the timing of yearly rainfall or temperature in the region and thereby challenged the mighty *Euphorbia*. Both the camel thorn and the quiver tree are also dying. "Trees with the least amount of flexibility seem to be going out first. They are tipping over the edge. It gives us an idea of how the more resilient ecosystems that humans depend on most might behave as the margins are pushed."

NO ONE knows the enigmatic world of trees better than Diana Beresford-Kroeger, an Irish-born botanist and self-defined "renegade scientist." Where most people see money and board feet, Beresford-Kroeger sees a living miracle, a chemical factory, a medicine cabinet, an oxygen tent, and ample food for the table. Outside of Merrickville, Ontario, the sixty-four-year-old scientist cultivates a sprawling garden with a hundred rare or disease-resistant trees including chestnuts, elms, Siberian cherries, and rare firs.

Beresford-Kroeger, who has the voice of an Irish songbird, is the kind of person who can tell you that the inner bark of many pines is so rich in carbohydrates, dietary fiber, and vitamin C that the Sami of Norway and the Gitksan of British Columbia regularly harvested it for food and ate the cambium fresh or roasted. She will tell you that trees once covered half the earth and now shade less than a third of the planet, yet still store about 45 percent of the world's carbon.

Beresford-Kroeger takes a broad and deep view of life. The great insect biologist E.O. Wilson thinks her ideas about reforesting cities and rural areas with trees that serve as living pharmacopeias, famine fighters, pollution absorbers, air fresheners, and insect controllers constitute "an entirely new approach to natural history." Beresford-Kroeger calls it bio-planning. It's simply the practice of planting seedlings from the oldest and healthiest native trees and arranging those trees on the landscape to perform a variety of good communal tasks. Conifers, for example, should be planted around schools and hospitals because of their restorative powers. Bio-planning is basically about noticing the connections between things, she adds.

Beresford-Kroeger would be the first to admit that she is not an expert on bark beetles. But she knows that foresters don't like the animal because beetles make mincemeat of human economic plans and arrogance. For starters, she says, removing the prospect of fire from the forest was a bad idea. If you remove one engine of renewal in a forest, another will take its place. "Nobody wants to look at the forest holistically," she says. "The great hand of Nature is something none of us understands."

Beresford-Kroeger says that conifers probably came into this world nearly 600 million years ago and that we should be grateful for their medicinal presence. Conifers departed the company of ferns and mosses to become their own clan of powerful carbon castles. They have an unrivaled capacity in the tree world to withstand drought, she says. They can photosynthesize at low temperatures and low light conditions as well as coat their seeds with antifungal and antibacterial chemicals.

In many ways, she says, the conifers changed the earth. They "sucked up the carbon dioxide and added back "the spit and polish of oxygen." Together with the oxygen-breathing plankton of the ocean, they helped make the world green and livable. Without the millions of multi-micron-sized mouths that every conifer manages, the first placenta-carrying mammals would never have had enough oxygen to make a go of it 60 million years ago.

With our plowing, tree felling, and carbon burning, human beings produce the carbon dioxide that the tree desires, and the tree produces the oxygen that humans needs. "The conifers changed the atmosphere," says Beresford-Kroeger. "But now we're going backward. We are going back to a time when CO_2 levels are high." In response, many conifers will pair off and die, she says; "they will go down." Others will hybridize. There will be new "sports" (the end result of epigenetic change) and new kinds of trees. "When you take them to the margin, they hybridize."

There might be other unexpected consequences, too, especially for mammals. Since prehistoric times, the concentration of oxygen in the atmosphere has dropped from 35 to 21 percent as the carbon dioxide created by humans rises. In polluted cities, oxygen often falls to only 15 percent of the mix. Both the world's abused forests and our polluted oceans are producing only half of the oxygen they did ten thousand years ago. As Beresford-Kroeger points out, a pregnant woman usually carries the fetus in her womb for between twenty-seven and thirty-seven weeks. During that time, the placenta provides food and oxygen to the fetus. If ambient oxygen levels drop further, women will need to carry their infants for longer periods of time, as many mountain dwellers already do. But for many women, thirty-seven weeks is already a dangerous stretch. "It's all a matter of molecular physics," says Beresford-Kroeger. "When the forests go down, women will suffer."

Beresford-Kroeger thinks the scientists and artists studying beetle sounds and the acoustics of trees are on to something very important. She says people register the infrasound produced by trees and forests "as an intense emotional experience."

(Infrasound is lower in frequency than twenty hertz per second, the usual limit of human hearing.) She likens the experience to the feelings people have when listening to great music. "The sounds, both audible and inaudible, of the pine, are sharper when compared to the rounder sounds of a red oak. The movement of the air as it travels through and with the pine is more finely dissected by pine needles to make the sharp whine, while the red oak leaves are more like the flapping of sails on a yacht." Trees in a forest probably communicate with each other using infrasound, too. Beresford-Kroeger would not be surprised by bark beetles hearing the weeping of drought-stressed trees.

The botanist also suspects that fungi are playing a hidden role in the beetle outbreaks. Every tree in the world has around its base a community of mycorrhizal fungi that fit over its roots like a glove. Most pines cannot be successfully planted without a good inoculation of the mycorrhizal assistants on their roots. These fungi help the tree uptake phosphorous as well as store carbon in the soil. But as atmospheric CO_2 increases, nobody knows how these diverse and complex fungal communities will respond. "All biochemical reactions are governed by temperature," says Beresford-Kroeger. Lower temperatures slow things down, whereas higher temperatures speed things up. "When you increase chemical reaction and rate, fungi will multiply."

Trees also harbor their own fungi, and warmer temperatures likely increase the fungal activity there, too. Beresford-Kroeger suspects the beetle has read this change, though she doesn't know for sure, since "we have not looked at the tree in a holistic manner." It's possible that increased levels of CO_2 may embolden these fungal communities or make them more capricious, but again, "nobody really knows." In her own garden the botanist has witnessed an explosive growth of sphagnum moss. "They are growing gangbusters." When the world had high CO_2 levels, she points out, these "tree-like structures" grew as large as conifers.

Several British studies suggest that fungal activity in general has grown exponentially with warming weather. Fungi are fruiting not only earlier but more often. One observant stonemason who's studied fungi as a hobby for over fifty years has discovered that nearly one-third of the species in Great Britain have started to fruit twice a year. In the United States, a native fungus once thought to be rare has multiplied so successfully with warming temperatures that it now threatens millions of Douglas-fir trees in Oregon. The fungus simply grinds the tree's growth to a halt. And a distinctly new race of *Puccinia graminis*, perhaps the world's most dangerous fungus, now threatens 90 percent of the world's wheat crop. Most wheat fields are as genetically uniform as a forest plantation.

Beresford-Kroeger does not know what the death of more than 30 billion trees in the West portends for the forest or for the region. However, she suspects that "we will be left with less diversity, and less diversity is not a good thing."

She rhymes off the affected species as a mother might account for the idiosyncrasies of her children. *Pinus contorta*, the lodgepole, a playful conifer and a lean water user, she says, will struggle. Despite its weedy abundance, the lodgepole is not the most vigorous of pines, and it resembles in some ways very blond or red-haired people. "Genetically, they are on the dicey side. They bleed more and have more cancer." In contrast, the *Pinus albicaulis*, the whitebark pine, is a stoic and hermit with the ability to withstand all sorts of adversity. "It's like a person who doesn't get diabetes," she says. Although its range and population will seriously dwindle, Beresford-Kroeger thinks the whitebark will remain a survivor. The Sitka spruce, on the other hand, is "a fat factory. As the climate forces it to go lean and mean, it will have problems." *Pinus edulis*, the pinyon, "a marvel of the evergreen world," spends enormous resources to make its civilization-building nuts and therefore is vulnerable. Losing the pinyon, "the jewel of the pine world," in the global forest would be like losing the Hopi or the Navajo in the global community, she laments. "We will have lost something very big."

Beresford-Kroeger characterizes the white spruce that dominate the boreal forest, *Picea glauca*, as flexible climate gymnasts. She calls them the most valuable and fundamental trees in the world, because they "can sense what's happening and have the genetic resources to protect themselves." Because of the white spruce's resilience, she doesn't think Canadians or Russians should be logging northern forests anymore.

In the end, Beresford-Kroeger regards the riotous beetle outbreaks in drought-stressed western forests as another illustration of how far humans have separated themselves from "the pulse of nature." More than half of the world's population lives in cities, and many urban dwellers can't identify a single tree. Civilization, she writes in her book *The Global Forest*, "has ignored the pattern in which nature works, as if man himself is an independent species apart from the web of it. The truth is that man is only one species, and he stands on a fragile platform of life that is but a whisper away from death."

TEN

The Parable of the Worm

. . . . .

"Nature which governs the whole will soon change all things which thou seest, and out of their substance will make other things, and again other things from the substance of them, in order that the world may be ever new."

MARCUS AURELIUS, *Meditations* (ca. 167 ACE)

TO UNDERSTAND the meaning of the bark beetle's determined reworking of western forests, we need to appreciate the parable of the eastern budworm (*Choristoneura fumiferana*). It may seem odd for a book about beetles to end with a tale about worms, but neither creature gives a damn about linear thinking.

Beetles and budworms erupt unpredictably, change everything, humiliate experts, and then abandon the scene of destruction like drunken prophets. The venerable insects work together in the forest renewal business, though they chew on different tree species. A western relative of the budworm, for example, has pretty much decimated the Douglas-fir in Alberta's Porcupine Hills and much Engelmann spruce in Montana. Wherever the tree-eating empire of *Choristoneura fumiferana* has collided with our own imperial behavior, be it logging or conservation ambitions, the fallout has been traumatic. The budworm has revealed something fundamental about how things can fall apart. At least that's what the eastern spruce budworm taught Crawford S. Holling nearly forty years ago.

Holling, perhaps the world's greatest living ecologist, is an eighty-year-old white-haired grandfather who lives in Nanaimo, British Columbia. Economists, foresters, historians, Internet gurus, and social critics quote his work with gusto. Almost every North American beetle expert is aware of Holling, popularly known as Buzz, or of his work. Alaskan scientists pepper their beetle papers with references to Holling and the importance of his thinking on "resilience." Allan Carroll wanted to study with Holling; Jesse Logan has a copy of Holling's *Panarchy* on his desk; and the musician David Dunn once spent a day with Holling and James Lovelock visiting the Uffizi gallery in Florence, Italy.

Holling's critical work about how small things can rapidly undo big, complex systems now pops up in the press as commonly as stories about corrupt financial institutions. But for more than half a century, the Toronto-born scientist worked largely in obscurity. He began by elucidating the shoddy management of fisheries, forests, and wetlands across North America. About three decades ago, he tabled some startling conclusions about why both forests and societies collapse. He noted that humans prize constancy in their economic affairs, whereas Mother Nature prefers variety. When these discordant ambitions clash, the collision, often silent or invisible, creates surprises and reveals the vulnerability of all the creatures involved.

Humans need to cultivate "owlish ways to hear the rustle of the mouse," Holling suggests. When he received the 2008 Volvo Environment Prize, he told his audience, "One has to learn to develop senses that help us listen to intriguing voices that are hidden amongst the noise." Unlike many experts today, Holling regards bark beetles and budworms as having serious and important voices. He also reckons that the extreme, the small, and the improbable will decide our future.

Holling is not alone in believing that marginal surprises can undo static central enterprises. Nassim Nicholas Taleb, the great economic critic and philosopher, reached exactly the same conclusions about the "impact of the highly improbable." He famously called these events "black swans." After studying globalization and economic concentration, Taleb recognized that rare and improbable events make history jump and empires crash. Before the big economic fallout of 2008, he warned that serious trouble was brewing. The world's banking system had grown so large and so interconnected that the failure of only one bank could trigger total financial collapse. The previous world ecology of small banks boasted varied lending practices and could accommodate frequent failure. But the experts transformed that robust ecosystem into a giant homogenous framework (much like a mature lodgepole forest) deemed "too big to fail." In removing all flexibility from the system, the financial gurus had exposed ordinary people to catastrophic risk. "Globalization creates interlocking fragility," wrote Taleb, "while reducing volatility and giving the appearance of stability."

In a 2010 update to his international bestseller, *The Black Swan*, Taleb makes some very Holling-like observations about Mother Nature, as he refers to the natural world. First, Mother Nature favors redundancy and inefficiency, and she avoids all forms of naive optimization. Second, she "does not like anything too big," because big things just aren't robust. Unlike bankers or empire builders, she prudently confines her mistakes and miscalculations to limited areas. She never over leverages or goes into debt. And Mother Nature laughs at engineered utopias. "Reducing volatility and ordinary randomness increases exposure to Black Swans—it creates an artificial quiet," Taleb observes. He believes that our economic life should resemble our biological environment: "smaller companies, richer ecology, and no leverage."

These days, Buzz Holling thinks a lot about resilience, which Taleb calls "robustness." The ecologist defines the term as the ability to recover from wild shocks or the unexpected, such as a storm of bark beetles. Nature is based on resilience, but humans have forgotten its value, says Holling. A healthy forest can absorb many surprises, from fires to insects, because its diverse membership ensures robustness over centuries. In contrast, humans generally design their economic and political affairs as big, dull, short-term monocultures without considering change, black swans, or disturbance as an immutable part of life. That's why Holling helped to found the Resilience Alliance, a research group that explores "the dynamics of complex adaptive systems." Ten-year-olds understand the mechanics of overcoming adversity much better than do fifty-year-olds, says Holling. You don't have to explain resilience to a child, but most professionals require jargon-filled textbooks on the subject.

In many ways, Holling, a graceful man, has an ancient view of nature. Unlike some environmentalists, the ecologist doesn't think the natural world operates like some fun-filled Disneyland. In Holling's universe, tragedy, wonder, poetry, loss, and instability punctuate natural life, with its unpredictable oscillations. It's doubtful such a dynamic place was ever in balance, or could be. Nature enjoys living on the edge, much like an extreme sports enthusiast. Humans generally have no idea where that edge lies, and they don't want to admit their ignorance, says Holling. Our attempts to engineer stability usually end with both the natural world and the human one tipping too far. Earth-shaking change tends to happen as quickly as a wolf kill, he believes. He sometimes uses the analogy of a raft on water that suddenly flips over, soaking the occupants. A lake, a fishery, a forest, and even the climate can flip over just as fast. In that respect, an aging forest dominated by mature spruce, pine, or fir, says Holling, is no different from a global financial system dominated by a few huge banks or a rigidly controlled political party. "It's an accident waiting to happen."

IT WAS in the 1970s that Buzz Holling, then a zoologist with the Canadian Forest Service, started to study the so-called budworm problem. At the time, the insect was devouring vast tracts of fir and spruce trees slated to become newsprint or toilet paper. Nearly three decades of intense logging and costly insecticide spraying with DDT had only repressed, not eradicated, the budworm. In fact, the indefatigable insect appeared to be infesting a much larger area of the forest than ever before. In a major 1910 outbreak, the budworm had consumed 10 million acres of trees. It now threatened to destroy more than 50 million acres. The pulp mills of New England and eastern Canada—the world's largest concentration—lay in its path. The emergency prompted New Brunswick chief forester Gordon Baskerville to invite Holling and a number of experts, including systems mathematicians, bug experts, and operations researchers, to analyze the whole damn thing. As Holling says, the project ended up being "one of the most detailed and exhaustive empirical studies of an ecosystem that has ever been attempted—the spruce budworm/forest interaction in eastern North America."

The budworm, like the bark beetle, is a remarkable silviculturalist and landscape engineer. Every forty years or so, the worm emerges from obscurity to explode in unbelievable numbers. These explosions generally kill aging stands of balsam fir, which most people recognize as a fine Christmas tree. The budworm doesn't kill quickly, like a bark beetle, but determinedly munches all the buds and needles over a decade. For the fir tree, it's a slow death by defoliation. The killing eventually allows new generations of birch and spruce to fill in the gaps and restore diversity in the forest. Seen from one angle, the budworm works like a visible forest competition bureau: it collapses monopolies of balsam, allocates space for spruce and birch, and then redistributes all the accumulated·capital.

Massive budworm outbreaks didn't garner too much attention until the beginning of the twentieth century, when the forest industry started to target fir for its long fibers and fine pulping qualities. About the same time (1910–1920) the budworm began one of its extraordinary population booms, devouring 27 million cords in the state of Maine—enough wood, as one analyst later calculated, to feed the state's 1980 pulp mill infrastructure for six years. Given that budworm outbreaks in Ontario and Quebec have extended over an area as large as 70,000 square miles (the size of Washington state), foresters have adopted a "kill-the-bug philosophy." In fact, the thinking behind eastern budworm wars later informed combat with western bark beetles.

By the 1960s, the budworm was one of the most studied and feared insect bogeymen in North America. Foresters typically defined the bug as "the most widespread and destructive defoliator of coniferous forests." Newspaper editors called it a "nasty little insect with a voracious appetite." On its website, the U.S. Department of Agriculture still refers to the budworm as "one of the most destructive native insects in the northern spruce and fir forests of the Eastern United States and Canada."

But Buzz Holling quickly realized that the budworm had no hidden agenda. It simply wanted to assure a continued food supply for itself by renewing the forest every thirty or forty years. The life cycle of the insect seems primed for creating a population boom at just the right moment for one great fir feast. The basic story goes like this. In July and August, clouds of moths appear and lay as many as ten clumps of twenty eggs each on the flat needles of a fir tree. Tiny larva hatch in about ten days and rappel down the tree on little silken threads. They shelter under bark over the winter. In late spring, these small larvae will molt several times as they become caterpillars. By the time the insect reaches a half-inch in length, it can chew an enormous volume of buds, needles, and cones. By midsummer, the worm pupates into a well-camouflaged moth, and the cycle begins again.

During a budworm boom, the moths frequently make aviation history. They take off in massive congregations from the forest canopy in order to catch wind gusts, which typically disperse the insects as far as 200 miles away. During one summer thunderstorm in 1974, scientists in a DC-3 saw a remarkable sight: a dense cloud of moths that occupied several thousand square miles of airspace about 1,600 feet above the ground. The moths, a huge blot on the radar, represented an airborne biomass of several hundred tons. It was like encountering a pod of twenty killer whales in the air.

During an outbreak, a budworm population can grow from a state of Mongolian modesty to Shanghai-like extremes. Most of the time, the budworm is a rare species. Scientists usually find no more than 5 caterpillars per tree. But during an outbreak, numbers can surge to hundreds of larvae per branch, about 22,000 per tree, in just four years. In full boom mode, the crowd of caterpillars on a balsam fir looks as ebullient as a stadium full of soccer fans. A spruce forest under budworm attack typically turns a dull rust color in the summer and assumes a deadly gray pallor by the fall. In three to six years of defoliation, the budworm can kill more than half of the forest. By the end of a 12-year-long feeding frenzy, almost every mature fir will be dead. At that point, the budworm crashes into obscurity. But scientists predict that climate change could increase the length of budworm outbreaks by 6 years and increase the scale of defoliation by 15 percent.

One of the first things the budworm taught Buzz Holling was that change is not continuous or gradual. At the time his work with the insect began, most ecologists assumed the world was globally stable, and that no matter where you lived, nature moved in a more or less linear fashion to some benign condition. The budworm proved that theory to be bullshit.

The boreal forest, like every ecosystem, moves through a series of stages: it grows, consolidates, and then collapses. Pioneer species such as birch and spruce start things off. Balsam fir crowds them out as it becomes the dominant player in the forest. As the fir matures over several decades, it consolidates energy and nutrition. But then the tree reaches a completely unstable phase. It loses its flexibility and resilience, and becomes rigid and complex. That's when fire or the budworm come along and change everything. During this destructive phase, "uncertainty is high, control is weakened and confused, and unpredictability is great." Without massive instability, however, the forest can't restore stability.

The budworm also taught Holling about the magical interplay between fast and slow and big and small in every forest. The aromatic balsam fir, which can live for up to 150 years, takes 7 to 10 years to grow back needles chomped by budworms. The budworm lives only a few months but can increase its numbers fivefold in one summer.

Birds play a critical role in keeping the budworm in check. Holling discovered that more than thirty-five different species of birds—the winter residents of tropical jungles, no less—dined on the worm, as did wasps and viruses. Predators such as the Tennessee warbler behaved like budworm gourmets. During an outbreak, the number of Tennessee warblers hunting in worm-filled forest climbs from zero to fifty pairs per ten-acre patch. The number of bay-breasted warblers expands from one to more than eighty pairs. "So long as the forest is immature, they keep the budworm population at low levels," says Holling. But once the foliage on a fir tree gets too dense and imposing, the birds lose interest in the budworm. With the help of dry, hot summers, the budworm takes off, and at this point the forest collapses. Holling calls it the revolt of the small against the slow and the big. Humans rarely pay enough attention to the slow variables, he says.

Perhaps the final lesson Holling gleaned from the budworm was the pathology of resource management. At the time the ecologist came on the scene, New Brunswick forestry officials had a rigid policy of "Keeping the Forest Green." (Maine had the same program.) The pulp industry regarded stands of graying or rusting trees as blights to commerce. So beginning in the 1950s, industry seized upon the miracle bug-killer DDT to kill budworms. (It later switched to fenitrothion, a nerve gas–like compound toxic to bees and harmful to fish and other wildlife.) Every spring, experts would map out a budworm war zone, and then dispatch a contingent of 220 planes over the forest. The small air force annually doused the forest with more than 3 million gallons of DDT or other chemical weapons.

Despite protests about rising costs, dead salmon, and rare illnesses in children, the spraying continued. "The actual politics of spraying had nothing to do with biology of the budworm," says Holling. A senior forestry executive ignored the side effects and successfully pushed for more spraying "because he was wealthy and personable. It was all driven by the short-term needs of industry."

Yet the spraying had become counterproductive. For about two decades, it temporarily delayed some needle-stripping infestations and tree deaths. But only that. Governments, however, assuming that DDT had conquered the worm, allocated more trees to pulp mills. The spraying maintained a mirage of budworm stability, says Holling. Yet by suppressing outbreaks, the annual spraying expanded the volume of mature fir on the landscape, the worm's ideal dinner. In reality, industry and government pesticide programs created "persistent semi-outbreak conditions" in which "more and more mature stands over larger areas were gradually deteriorating from the pressure of moderate but persistent budworm defoliation." Spraying created a highly vulnerable forest the same way fire suppression in the West created aging stands of pine.

When human engineers manage a forest solely to achieve a constant production of trees, the forest loses its resilience. As Holling puts it, "Short-term success in stabilizing production leads to long-term surprise." The institutions designed to manage the forest tend to become myopic and rigid over time, ignoring evidence that their decisions are causing harm and wasting money. The industries profiting from this pathology become fat and lazy, since engineered fantasies of constant and efficient production don't nourish vision or innovation. The public loses trust in the whole corrupt process as the professionals in charge become increasingly aloof and secretive. Nothing fundamental changes until a crisis "triggered by unexpected external events" flips the whole arrangement.

The mathematical modeling performed by Holling and his colleagues eventually led to a radical report on forest conditions. First they showed that spraying not only made the problem worse but also was uneconomic. They also demonstrated that available stocks of harvestable trees were rapidly declining. In Holling's words, "industry's math was fundamentally flawed." Their report, entitled *A Case Study of Forest Ecosystem/Pest Management*, provoked an overhaul of forestry management in parts of eastern Canada.

The bluntly worded report still hums with an uncomfortable relevance. "Past efforts in resource management have been essentially trial and error approaches to coping with the unknown. And indeed that is the way our society has advanced since the industrial revolution. Existing information is mobilized to suggest a trial and if an error is detected then that provides additional information to modify subsequent trials. But we are now at the point where the intensity and extensiveness of our trials generate errors that are potentially larger than our society can afford. Trial and error seems increasingly to be a dangerous method for coping with the unknown. We need a new strategy to deal with ignorance." If governments and industry didn't change both their short-term and their big-scale approaches to land and water management, the report said, then the result would be "larger disasters, achieved faster and in a more pretentious and disciplined manner." It was a classic Taleb-like pronouncement long before Taleb.

BUZZ HOLLING thinks the bark beetle is pretty much repeating the budworm's message about collapse and renewal. The great beetle epidemics, he says, are no mystery, because human engineers created them. By suppressing fire in lodgepole and ponderosa forests throughout the West, we took a patchy and diverse forest with trees of all ages and turned it into a boring, middle-aged green mall. For more than twenty years by now, research has shown that the removal of fire hazards from a ponderosa forest, for example, increases tree density tenfold. At the same time, tree diameter decreases from an average of seventeen inches to ten inches. A forest deprived of fire and the resulting opportunity for renewal becomes a collection of toothpicks less resistant to drought and more vulnerable to insects and disease. Determined fire suppression *guarantees*, sooner or later, either catastrophic fire or imperial legions of bark beetles.

Westerners also overleveraged the forest system by either leasing too much land to forestry companies or creating parks with no respect for the rules of natural disturbance. And the trigger for the crisis was climate change. Increased temperatures allowed the beetle and its fungal allies to cross into new frontiers and transgress old boundaries.

Holling views the beetle outbreak as a novel opportunity for us to come to terms with our ignorance and accept the importance of robustness. But he doesn't think politicians or forestry corporations have absorbed any critical lessons yet. As a consequence, he suspects, whatever is large, global, and concentrated will not survive in the years ahead. What is small, flexible, and locally based will grow like lodgepole seedlings after a fire.

The downfall of western forests, Holling warns, could presage the collapse of complex fossil-fuel energy networks, the global financial system, and overleveraged global food supplies. "The forest makes a comprehensive metaphor, and it should be used to understand other phenomena on the planet. Now is a time of great and extensive turbulence." The world is facing major collapses on a scale never before experienced, and "all of these collapses have the same characteristics as the beetle and the lodgepole forest. We are ignoring the slow changes that are accumulating under our noses." Everywhere Holling looks, he says, he sees "efficient and expanded growth" that has carelessly "accumulated excessive opaqueness and rigidities." Economists and managers conjure up temporary solutions that "hide problems as they try to correct them, to finally create an accident waiting to happen."

The beetle has spoken, says Holling. "The future is not just uncertain. It is inherently unpredictable." So the choice is ours. Humans can collapse like an exhausted empire of *Dendroctonus* or renew ourselves like a seedling in among the deadfall.

"We create dangerous times full of surprises," concludes Holling. "But those surprises contain opportunities. Individuals and small groups can discover them, for they are the seeds of tomorrow."

Sources and Further Reading

. . . . .

General

Beresford-Kroeger, Diana. *Arboretum America: A Philosophy of the Forest.*
    Ann Arbor, MI: University of Michigan Press, 2003.

———. *The Global Forest.* New York: Viking, 2010.

Elton, Charles. *Animal Ecology.* New York: Macmillan, 1927.

———. *The Pattern of Animal Communities.* New York: John Wiley
    and Sons, 1966.

———. *Voles, Mice and Lemmings: Problems in Population Dynamics.* Oxford:
    Clarendon Press, 1942.

Evans, Arthur, and Charles Bellamy. *An Inordinate Fondness for Beetles.*
    New York: Henry Holt, 1996.

Evans, Glyn. *The Life of Beetles.* New York: Hafner Press, 1975.

Kritsky, Gene, and Ron Cherry. *Insect Mythology.* San Jose, CA:
    Writers Club Press, 2000.

Lockwood, Jeffrey. *Locust: The Devastating Rise and Mysterious
    Disappearance of the Insect That Changed the American Frontier.*
    New York: Basic Books, 2004.

————. "Voices from the Past: What We Can Learn from the Rocky Mountain Locust." *American Entomologist* 47 (2001).

Perlin, John. *A Forest Journey: The Role of Wood in the Development of Civilization.* Cambridge, MA: Harvard University Press, 1991.

Pyne, Stephen. *Awful Splendour: A Fire History of Canada.* Vancouver: UBC Press, 2007.

Schowalter, Timothy. *Insect Ecology: An Ecosystem Approach.* San Diego, CA: Academic Press, 2000.

Swaine, James M. *Canadian Bark Beetles.* Bulletin No. 14. Division of Entomology, Department of Agriculture, Dominion of Canada, 1917–18.

Taleb, Nassim Nicholas. *The Black Swan: The Impact of the Highly Improbable.* 2nd edition. New York: Random House, 2010.

Tomback, Diana, Stephen Arno, and Robert Keane, eds. *Whitebark Pine Communities: Ecology and Restoration.* Washington, D.C.: Island Press, 2001.

U.S. Department of the Interior. *First Annual Report of the United States Entomological Commission for the Year 1877, Relating to the Rocky Mountain Locust.* Washington, D.C.: Government Printing Office, 1878.

Williams, Michael. *Deforesting the Earth: From Prehistory to Global Crisis.* Chicago: University of Chicago Press, 2003.

One: The Alaska Storm

Bakke, Alf. "The Recent *Ips typographus* Outbreak in Norway: Experiences from a Control Program." *Holarctic Ecology* 12 (1989).

Berg, Ed. Contributions to *Refuge Notebook.* Kenai National Wildlife Refuge, U.S. Fish & Wildlife Service, 1998–2009. Available at http://kenai.fws.gov/notebook.htm.

Berg, Edward, J. David Henry, Christopher L. Fastie, Andrew D. De Volder, and Steven M. Matsuoka. "Spruce Beetle Outbreaks on the Kenai Peninsula, Alaska and Kluane National Park and Reserve, Yukon Territory: Relationship to Summer Temperatures and

Regional Differences in Disturbance Regimes." *Forest Ecology and Management* 227 (2006).

Berg, Edward, Kacy McDonnell Hillman, Roman Dial, and Allana DeRuwe. "Recent Woody Invasion of Wetlands on the Kenai Peninsula Lowlands, South-Central Alaska: A Major Regime Shift after 18 000 Years of Wet Sphagnum Sedge Peat Recruitment." *Canadian Journal of Forest Research* 39 (2009).

Cozens, Russ. "The Upper Bowron Spruce Beetle Outbreak: A Case History." Forest Entomology Textbook Challenge for the 21st Century. August 2004. Available at www.forestry.ubc.ca/fetch21/Upper%20Bowron%20Spruce%20Beetle/ubsbo.htm.

Egan, Timothy. "Alaska, No Longer So Frigid, Starts to Crack, Burn and Sag." *New York Times*, 16 June 2002.

Flint, Courtney. "Community Perspectives on Spruce Beetle Impacts on the Kenai Peninsula, Alaska." *Forest Ecology and Management* 227 (2006).

Garbutt, Rod, Brad Hawkes, and Eric Allen. *Spruce Beetle and the Forests of the Southwest Yukon*. Information Report BC-X-406. Ottawa: Natural Resources Canada, 2006.

Heinrichs, Jay. "Camp Apocalypse." *Backpacker*, June 2003.

Henry, David, Anne Landry, Tom Elliot, Laura Gorecki, Michael Gates, and Channy Chow. *State of the Park Report: Kluane National Park and Reserve of Canada*. Ottawa: Parks Canada, April 2008.

Juday, Glenn. "Spruce Beetles, Budworms and Climate Warming." *Global Glimpses* 6:1 (1998). Available at www.cgc.uaf.edu/newsletter/gg6_1/beetles.html.

Kenai Peninsula Borough, Spruce Bark Beetle Mitigation Program. *Annual Report* CY 2007. [2008.] Available at www.borough.kenai.ak.us/sbb/documents/exec_sum/07annualreport.pdf.

Lange, Holger, Bjørn Økland, and Paal Krokene. "Thresholds in the Life Cycle of the Spruce Bark Beetle under Climate Change." The Norwegian Forest and Landscape Institute, 2006. Available at www.necsi.edu/events/iccs6/papers/60bbc513f666a7a1677d01e4e3a0.pdf.

Matsuoka, S.M., E.H. Holsten, R.A. Werner, and R.E. Burnside. "Spruce
Beetles and Forest Ecosystems in South-Central Alaska: A Review
of 30 Years of Research." *Forest Ecology and Management* 227 (2006).

McCarty, Marie. "Elegy to a Spruce Forest: A Cemetery of Sorts."
Homer News, 17 April 2003.

Pretzlaw, Troy, Caroline Trudeau, Murray M. Humphries, Jalene
M. LaMontagne, and Stan Boutin. "Red Squirrels (*Tamiasciurus
hudsonicus*) Feeding on Spruce Bark Beetles (*Dentroctonus rufipennis*):
Energetic and Ecological Implications." *Journal of Mammalogy*
87 (2006).

Schroeder, Lief Martin. "Escape in Space from Enemies: A Comparison
between Stands with and without Enhanced Densities of the Spruce
Bark Beetle." *Agricultural and Forest Entomology* 9 (2007).

Wermelinger, Beat. "Ecology and Management of the Spruce Bark Beetle
Ips typographus: A Review of Recent Research." *Forest Ecology and
Management* 202 (2004).

Two: The Beetle, the Bus, and the Carbon Castle

Berryman, Alan, Brian Dennis, Kenneth F. Raffa, and Nils Christian
Stenseth. "Evolution of Optimal Group Attack, with Particular
Reference to Bark Beetles (Coleoptera: *Scolytidae*)." *Ecology* 66 (1985).

Cardoza, Y.J., J.C. Moser, K.D. Kepzig, and K.F. Raffa. "Multipartite
Symbioses among Fungi, Mites, Nematodes and the Spruce Beetle,
Dendroctonus rufipennis." *Environmental Entomology* 37 (2008).

Coulson, Robert. "Population Dynamics of Bark Beetles." *Annual Review
of Entomology* 24 (1979).

Franceshi, V.R., P. Krokene, E. Christiansen, and T. Krekling.
"Anatomical and Chemical Defenses of Conifer Bark against Bark
Beetles and Other Pests." *New Phytologist* 167 (2005).

Gershenzon, Jonathan, and Natalia Dudareva. "The Function of Terpene
Natural Products in the Natural World." *Nature Chemical Biology* 3
(2007).

Klepzig, K.D., J.C. Moser, M.J. Lombardero, M.P. Ayres, R.W. Hofstetter,
and C.J. Walkinshaw. "Mutualism and Antagonism: Ecological

Interactions among Bark Beetles, Mites and Fungi." In *Biotic Interactions in Plant Pathogen Associations*, edited by M.J. Jeger and N.J. Spence. New York: CABI Publishing, 2001.

Moser, John, Heino Konrad, Stacy Blomquist, and Thomas Kirisitis. "Do Mites Phoretic on Elm Bark Beetles Contribute to the Transmission of Dutch Elm Disease?" *Naturwissenshaften* 97 (2010).

Mueller, Ulrich, and Nicole Gerardo. "Fungus-Farming Insects: Multiple Origins and Diverse Evolutionary Histories." PNAS 99 (26 November 2002).

Rudinsky, J.A. "Ecology of Scolytidae." *Annual Review of Entomology* 7 (1962).

Scott, Jarrod J., Dong-Chan Oh, M. Cetin Yuceer, Kier D. Klepzig, Jon Clardy, and Cameron R. Currie. "Bacterial Protection of Beetle-Fungus Mutualism." *Science*, 3 October 2008.

Six, Diana, and Barbara Bentz. "Fungi Associated with the North American Spruce Beetle, *Dendroctonus rufipennis*." *Canadian Journal of Forest Research* 33 (2003).

Six, Diana, and Michael Wingfield. "The Role of Phytopathogenicity in Bark Beetle–Fungus Symbioses: A Challenge to a Classic Paradigm." *Annual Review of Entomology* 56 (2011).

Wood, Daniel. "The Role of Pheromones, Kaironmones, and Allomones in the Host Selection and Colonization Behavior of Bark Beetles." *American Review of Entomology* 27 (1982).

Wood, Stephen L. *The Bark and Ambrosia Beetles of North and Central America (Coleoptera: Scolytidae): A Taxonomic Monograph*. Great Basin Naturalist Memoirs 6. Provo, UT: Brigham Young University, 1982.

Three: The Lodgepole Tsunami

Aukema, Brian H., Richard A. Werner, Kirsten E. Haberkern, Barbara L. Illman, Murray K. Clayton, and Kenneth F. Raffa. "Quantifying Sources of Variation in the Frequency of Fungi Associated with Spruce Beetles: Implications for Hypothesis Testing and Sampling Methodology in Bark Beetle–Symbiont Relationships." *Forest Ecology and Management* 217 (2005).

Carroll, Allan, Steve W. Taylor, Jacques Régnière, and Les Safranyik.
"Effects of Climate Change on Range Expansion by the Mountain
Pine Beetle in British Columbia." Presented at the Mountain Pine
Beetle Symposium: Challenges and Solutions, Kelowna, B.C.,
October 2003. Available at www4.nau.edu/direnet/publications/
publications_c/files/Carrol_et_al_2003.pdf.

Forest Practices Board. "Nadina Beetle Treatments: Complaint
Investigation 030500." FRB/IRC/99. Victoria, B.C.: Forest
Practices Board, November 2004.

Forestry Canada. *Annual Report of the Forest Insect and Disease Survey.*
Ottawa: Forestry Canada, 1936–96.

Gryse de, J.J., and A.W.A. Brown. "Co-operative Forest
Insect Survey." Presented at the Annual Meeting of the
Woodlands Section, Canadian Pulp and Paper Association,
January 1939.

Jackson, Peter L., Dennis Straussfogel, B. Staffan Lindgren, Selina
Mitchell, and Brendan Murphy. "Radar Observation and Aerial
Capture of Mountain Pine Beetle, *Dendroctonus ponderosae* Hopk.
(Coleoptera: *Scolytidae*) in Flight above the Forest Canopy (Report)."
Canadian Journal of Forest Research 38 (2008).

Koch, Peter. *Gross Characteristics of Lodgepole Pine Trees in North America.*
General Technical Report INT-227. Forest Service, United States
Department of Agriculture, 1987.

Machmer, Marlene, and Christoph Steeger. *The Ecological Roles of
Wildlife Tree Users in Forest Ecosystems.* Land Management Handbook.
Victoria: Ministry of Forests Research Program, Province of British
Columbia, 1995.

Nikiforuk, Andrew. "Pine Plague." *Canadian Geographic*, January/
February 2007.

Raffa, Kenneth, and Alan Berryman. "Interacting Selective Pressures in
Conifer–Bark Beetle Systems: A Basis for Reciprocal Adaptations?"
The American Naturalist 129 (1987).

Raffa, Kenneth F., Brian H. Aukema, Barbara J. Bentz, Allan L. Carroll,
Jeffrey A. Hicke, Monica G. Turner, and William H. Romme.

"Cross-scale Drivers of Natural Disturbances Prone to
Anthropogenic Amplification: The Dynamics of Bark Beetle
Eruptions." *BioScience* 58 (2008).

Safranyik, L., and A. Carroll. "The Biology and Epidemiology of the
Mountain Pine Beetle in Lodgepole Pine Forests." In *The Mountain
Pine Beetle: A Synthesis of Biology, Management, and Impacts on Lodgepole
Pine,* edited by L. Safranyik and W.R. Wilson. Victoria, B.C.: Pacific
Forestry Centre, Canadian Forest Service, Natural Resources
Canada, 2006.

Taylor, S.W., and A.L. Carroll. "Disturbance, Forest Age, and Mountain
Pine Beetle Outbreak Dynamics in B.C.: A Historical Perspective."
Presented at the Mountain Pine Beetle Symposium: Challenges
and Solutions, Kelowna, B.C., October 2003.

White, Patrick. "Red Rush." *The Walrus,* April 2007.

Wilson, John, Pat Bell, and Paul Nettleton. *Report of the Mountain Pine
Beetle Task Force.* Victoria: B.C. Minister of Forests, 3 October 2001.

Four: The War against the Insect Enemy

Black, Scott Hoffman. *Logging to Control Insects: The Science and Myths
behind Managing Forest Insect "Pests": A Synthesis of Independently
Reviewed Research.* Portland, OR: The Xerces Society for Invertebrate
Conservation, 2005. Available at www.xerces.org/wp-content/
uploads/2008/10/logging_to_control_insects.pdf.

Bláha, Jaromir. "Spruce Monocultures in the Czech Republic: The
Sumava Mountains Case Study." Prepared for the Sixth Conference
of the Parties of the Framework Convention on Climate Change.
Friends of the Earth International/World Rainforest Movement/
FERN, 2001. Available at www.wrm.org.uy/actors/CCC/
trouble5.html.

Borden, John H., Glen R. Sparrow, and Nicole L. Gervan. "Operational
Success of Verbenone against the Mountain Pine Beetle in a Rural
Community." *Arboriculture & Urban Forestry* 33 (2007).

Chamberlin, W.J. *The Bark and Timber Beetles of North America.* Corvallis,
OR: OSC Cooperative Association, 1939.

Furniss, Malcolm M. "American Forest Entomology Comes on Stage: Bark Beetle Depredations in the Black Hills Forest Reserve, ca. 1897–1907." *American Entomologist* 43 (1997).

Gibson, Ken. *Using Verbenone to Protect Host Trees from Mountain Pine Beetle Attack.* Missoula, MT: USDA Forest Service, January 2009.

Gmelin, Joh. Freidr. *Abhandlung über die Wurmtroknis* [Treatise on the Worm Dryness]. Liepzig: Verlay Crusious, 1787.

Hopkins, Andrew Delmar. *The Black Hills Beetle: With Further Notes on Its Distribution, Life History, and Methods of Control.* Washington, D.C.: U.S. Department of Agriculture, 1905.

———. *The Dying of Pine in the Southern States: Cause, Extent and Remedy.* Washington, D.C.: U.S. Department of Agriculture, 1911.

———. *Insect Enemies of the Spruce in the Northeast: A Popular Account of Results of Special Investigations, with Recommendations for Preventing Losses.* Washington, D.C.: U.S. Division of Entomology, 1901.

———. *Practical Information on the Scolytid Beetles of North American Forests. I. Bark Beetles of the Genus* Dendroctonus. Washington, D.C.: U.S. Department of Agriculture, 1909.

Jonasova, Magda, and Karel Prach. "Central European Mountain Spruce (*Picea abies*) Forests: Regeneration of Tree Species after a Bark Beetle Outbreak." *Ecological Engineering* 23 (2004).

Miller, J.M., and F.P. Keen. *Biology and Control of the Western Pine Beetle: A Summary of the First Fifty Years of Research.* Washington, D.C.: U.S. Department of Agriculture, March 1960.

Price, Terry S., Coleman Doggett, John M. Pye, and Bryan Smith, eds. *A History of Southern Pine Beetle Outbreaks in the Southeastern United States.* Southern Forest Insect Working Group, 2006. Available at www.barkbeetles.org/spb/index.html.

Proceedings of the Entomological Society of British Columbia, nos. 17/19; Economic Series, no. 24, Evans and Hastings, October 1923 and 1924.

Richmond, Hector. *Forever Green: The Story of One of Canada's Foremost Foresters.* Lanzville, B.C.: Oolichan Books, 1983.

Safranyik, Les, and H.S. Whitney. "Using Explosives to Destroy Mountain Pine Beetle Broods in Lodgepole Pine Trees." *Journal of the Entomological Society of British Columbia* 77 (1980).

Swaine, J.M. *Canadian Bark Beetles: Part II: A Preliminary Classification with an Account of the Habits and Means of Control.* Entomological Bulletin. Ottawa: Department of Agriculture, 1918.

———. *Forest Insect Conditions in British Columbia: A Preliminary Survey.* Ottawa: Department of Agriculture, 1914.

United Kingdom, Forestry Commission. *British Bark-Beetles.* Forestry Commission Bulletin No. 8. London: His Majesty's Stationery Office, 1926.

Vite, J.P. "The European Struggle to Control *Ips typographus*— Past, Present and Future." *Holarctic Ecology* 12 (1989).

Wickman, Boyd E. *The Battle against Bark Beetles in Crater Lake National Park: 1925–34.* General Technical Report PNW-GTW-259. Washington, D.C.: U.S. Department of Agriculture, June 1990.

Five: The Wake of the Beetle

Alberta Mountain Pine Beetle Advisory Committee. *Recommendations for Community Sustainability.* September 2007.

Alila, Younes, and Charles Luo. "Peak Flow and Water Yield Response to Clearcut Salvage Logging in Meso-scale Mountain Pine Beetle Infested Watersheds." Vancouver: University of British Columbia, 20 January 2007.

Alila, Younes, Piotr K. Kuraś, Markus Schnorbus, and Robert Hudson. "Forests and Floods: A New Paradigm Sheds Light on Age-Old Controversies." *Water Resources Research* 45 (2009).

Britneff, Anthony. "Neglect in the Woods: No Way to Manage a Forest." *Times Colonist* (Victoria, B.C.), 10 June 2010.

Brown, M., T.A. Black, Z. Nesic, V.N. Ford, D.L. Spittlehouse, A.L. Fredeen, N.J. Grant, P.J. Burton, and J.A. Trofymow. "Impact of Mountain Pine Beetle on the Net Ecosystem Production of Lodgepole Pine Stands in British Columbia." *Agriculture and Forest Meteorology* 150 (2009).

Forest Practices Board. *The Effect of Mountain Pine Beetle Attack and Salvage Harvesting on Streamflows: Special Investigation.* FPB/SIR/16. Victoria, B.C.: Forest Practices Board, March 2007.

———. *Species Composition and Regeneration in Cutblocks in Mountain Pine Beetle Areas.* FPB/SIR/15. Victoria, B.C.: Forest Practices Board, October 2006.

Georgeson, Annerose. *Hope Persists: Artwork from a Changed Forest.* Catalogue. Prince George, B.C.: Two Rivers Gallery, 2007.

———. *Red and Blue Beetle Art.* Catalogue. Prince George, B.C.: Omineca Beetle Action Coalition, 2008.

Huber, Dezene P.W., Brian H. Aukema, Robert S. Hodgkinson, and B. Staffan Lindgren. "Successful Colonization, Reproduction, and New Generation Emergence in Live Interior Hybrid Spruce *Picea engelmanniixglauca* by Mountain Pine Beetle *Dendroctonus ponderosae.*" *Agricultural and Forest Entomology* 11 (2009).

Jones, Jeff. "A Theology of Impotence." Sermon delivered at Calvary Grace Church of Calgary, 28 May 2010.

Konkin, Doug, and Kathy Hopkins. "Learning to Deal with Climate Change and Catastrophic Forest Disturbances." *Unasylva* (Rome: Food and Agricultural Organization of the United Nations) 60:231/232 (2009).

Kurz, Werner A., Graham Stinson, Gregory J. Rampley, Carn C. Dymond, and Eric T. Neilson. "Risk of Natural Disturbances Makes Future Contribution of Canada's Forests to the Global Carbon Cycle Highly Uncertain." *PNAS* 105 (2008).

Latty, Tanya, and Mary Reid. "Who Goes First? Condition and Danger Dependent Pioneering in a Group-Living Bark Beetle (*Dendroctonus ponderosae*)." *Behavioral Ecology Sociobiology* 64 (2010).

Mountain Pine Beetle Information Network. Available on FORREX at http://nrin.forrex.org/servlet/mpb.

Neads, Dave. "The Pine Beetle Aftermath Requires a Fundamental Culture Change." *Inside British Columbia: The Blog,* 3 July 2007.

Parfitt, Ben. *Battling the Beetle: Taking Action to Restore British Columbia's Interior Forests*. Vancouver: Canadian Centre for Policy Alternatives, July 2005.

———. *Managing BC's Forests for a Cooler Planet: Carbon Storage, Sustainable Jobs and Conservation*. Vancouver: Canadian Centre for Policy Alternatives, January 2010.

———. *Overcutting and Waste in B.C.'s Interior: A Call to Rethink B.C.'s Pine Beetle Logging Strategy*. Vancouver: Canadian Centre for Policy Alternatives, June 2007.

von Tiesenhausen, Peter. *Requiem*. Catalogue. Prince George, B.C.: Two Rivers Gallery, July 2005.

Williamson, T.B. *Assessing Potential Biophysical and Socioeconomic Impacts of Climate Change on Forest-Based Communities: A Methodological Case Study*. Information Report NOR-X-415E. Ottawa: Canadian Forest Service, Natural Resources Canada, 2008.

Six: The Ghost Forest

Bentz, Barbara, and Greta Schen-Langenheim. "The Mountain Pine Beetle and Whitebark Pine Waltz: Has the Music Changed?" In *Proceedings of the Conference Whitebark Pine: A Pacific Coast Perspective*, edited by E.M. Goheen and R.A. Sneizko. Portland, OR: Forest Service, Pacific Northwest Region, U. S. Department of Agriculture, 2007.

Cart, Julie. "Yellowstone a Petri Dish for Climate Change." *Los Angeles Times*, 6 December 2009.

Gibson, Ken, Kjerstin Skov, Sandy Kegley, Carl Jorgensen, Sheri Smith, and Jeff Witcosky. *Mountain Pine Beetle Impacts in High-Elevation Five-Needle Pines: Current Trends and Challenges*. RI-08-020. Washington, D.C.: Forest Health Protection, Forest Service, United States Department of Agriculture, September 2008.

Lanner, Ronald. *Made for Each Other: A Symbiosis of Birds and Pines*. Oxford: Oxford University Press, 1996.

Logan, Jesse, and James Powell. "Ecological Consequences of Climate
Change Altered Forest Insect Disturbance Regimes." In *Climate
Change in Western North America*, edited by F.H. Wagner. Salt Lake
City: University of Utah Press, 2005.

———. "Ghost Forests, Global Warming and the Mountain Pine Beetle
(Coleoptera: *Scolytidae*)." *American Entomologist* 47 (2001).

Logan, Jesse, P.V. Bolstad, B.J. Bentz, and D.L. Perkins. "Assessing the
Effects of Changing Climate on Mountain Pine Beetle Dynamics."
In *Interior West Global Climate Workshop* (Proceedings), edited by
R.W. Tinus. USDA Forest Service GTR-RM-262. Fort Collins, CO:
Rocky Mountain Forest and Range Experiment Station, Forest
Service, U.S. Department of Agriculture, 1995.

Logan, Jesse, William W. Macfarlane, and Louisa Willcox. "Whitebark
Pine Vulnerability to Climate-Driven Mountain Pine Beetle
Disturbance in the Greater Yellowstone Ecosystem." *Ecological
Applications* 20 (2010).

Macfarlane, W.W., J.A. Logan, and W.R. Kern. *Using the Landscape
Assessment System* (LAS) *to Assess Mountain Pine Beetle–Caused
Mortality of Whitebark Pine, Greater Yellowstone Ecosystem, 2009:
Project Report*. Prepared for the Greater Yellowstone Coordinating
Committee, Whitebark Pine Subcommittee, Jackson, Wyoming,
2010. GEO/Graphics. Available at http://docs.nrdc.org/land/files/
lan_10072101a.pdf.

Nijhuis, Michelle. "Global Warming's Unlikely Harbingers." *High
Country News*, 19 July 2004.

Riccardi, Nicholas. "Climate Blamed for Aspen Deaths." *Los Angeles
Times*, 18 October 2009.

Willcox, Louisa, "A Grizzly Future: The Loss of Whitebark Pine and
the Building of a Perfect Storm for Yellowstone's Grizzly Bears."
Natural Resources Defense Council, July 2009.

Seven: The Song of the Beetle

The Acoustic Ecology Institute. Acousticecology.org.

Adams, Henry D., Maite Guardiola-Claramonte, Greg A. Barron-
Gafford, Juan Camilo Villegas, David D. Breshears, Chris B. Zou,
Peter A. Troch, and Travis E. Huxman. "Temperature Sensitivity
of Drought-Induced Tree Mortality Portends Increased
Regional Die-Off under Global Change–Type Drought."
PNAS 106 (2009).

Allen, Craig, and David Breshears. "Drought-Induced Shift of a Forest-
Woodland Ecotone: Rapid Landscape Response to Climate
Variation." PNAS 95 (1998).

Breshears, David, et al. "Regional Vegetation Die-Off in Response to
Global-Change-Type Drought." PNAS 102 (2005).

Cochard, Hervé. "Cavitation in Trees." C.R. *Physique* 7 (2006).

Crutchfield, James. "What Lies between Order and Chaos?" Santa Fe
Institute, 11 March 2002.

Dunn, David. *The Sound of Light in Trees: Bark Beetles and the Acoustic
Ecology of Pinyon Pines.* CD. Earth Ear and Acoustic Ecology Institute,
2006.

Dunn, David, and James Crutchfield. "Entomogenic Climate Change:
Insect Bioacoustics and Future Forest Ecology." *Leonardo* 42 (2009).

Floyd, M. Lisa., Michael Clifford, Neil S. Cob, Dustin Hanna,
Robert Delph, Paulette Ford, and Dave Turner. "Relationship of
Stand Characteristics to Drought-Induced Mortality in Three
Southwestern Pinyon-Juniper Woodlands." *Ecological Applications*
19 (2009).

"Here Comes the Sound." *Living on Earth.* Podcast. 26 February 2010.
Download and transcript available at www.loe.org/shows/segments.
htm?programID=10-P13-00009&segmentID=6.

Huang, Cho-ying, Gregory P. Asner, Nichole N. Barger, Jason C. Neff,
and M. Lisa Floyd. "Regional Aboveground Live Carbon Losses
Due to Drought-Induced Tree Dieback in Piñon-Juniper
Ecosystems." *Remote Sensing of Environment* 114 (2010).

Lanner, Ronald M. *The Piñon Pine: A Natural and Cultural History.* Reno:
University of Nevada Press, 1981.

Pinon Nuts.org: A Site for Promoting the Sustainable Harvest of Wild Pinyon Nuts. www.pinonnuts.org.

Proceedings: Pinyon-Juniper Conference. General Technical Report INT-215. Washington, D.C.: Forest Service, U.S. Department of Agriculture, January 1987.

Raffles, Hugh. Insectopedia. New York: Pantheon Books, 2010.

Rudinsky, J.A., and R.R. Michael. "Sound Production in Scolytidae: 'Rivalry' Behaviour of Male Dendroctonus Beetles." Journal of Insect Physiology 20 (1974).

Eight: The Sheath-Winged Cosmos

Beutel, Rolf, Frank Friedrich, and Richard Leschen. "Charles Darwin, Beetles and Phylogenetics." Naturwissenschaften 96 (2009).

Cowan, Frank. Curious Facts in the History of Insects, Including Spiders and Scorpions. Philadelphia: J.B. Lippincott, 1865.

Dalziel, Louise. "Jewel Beetle Flies into the Inferno." BBC News, 20 March 2005. Available at http://news.bbc.co.uk/2/hi/science/nature/4362589.stm.

Erwin, Terry L. "Tropical Forests: Their Richness in Coleoptera and Other Arthropod Species." The Coleopterists Bulletin 36:1 (1982).

Evans, E.P. The Criminal Prosecution and Capital Punishment of Animals. London: W. Heinemann, 1906.

Fabre, J. Henri. The Sacred Beetle and Others. New York: Dodd, Mead and Company, 1918.

Farrell, Brian D., Andrea S. Sequeira, Brian C. O'Meara, Benjamin B. Normark, Jeffrey H. Chung, and Bjarte H. Jordal. "The Evolution of Agriculture in Beetles (Curculionidae: Scolytinae and Platypondinae)." Evolution 55 (2001).

Girgen, Jen. "The Historical and Contemporary Prosecution and Punishment of Animals." Animal Law 97 (2003).

Menéndez, Rosa. "How Are Insects Responding to Global Warming?" Tijdschrift voor Entomologie 150 (2007).

Peterson, Robert. "Charley Patton and His Mississippi Boweavil Blues." American Entomologist 53 (2007).

Ratcliffe, Brett C. *Scarab Beetles in Human Culture.* Papers in
   Entomology. Lincoln: Museum, University of Nebraska State,
   2006.
Robson, David. "Why a Big Horn Gives Beetles a Tiny Organ."
   The New Scientist 199 (3 September 2008).
Scholtz, Gerhard. "Scarab Beetles at the Interface of Wheel Invention
   in Nature and Culture?" *Contributions to Zoology* 77 (2008).
Wade, Nicholas. "Extravagant Results of Nature's Arms Race."
   New York Times, 23 March 2009.
Yoon, Carol Kaesak. "A Taste for Flowers Helped Beetles Conquer the
   World: In Evolutionary Biology, Diet Is Destiny." *New York Times,*
   28 July 1998. Available at www.nytimes.com/1998/07/28/science/
   taste-for-flowers-helped-beetles-conquer-world-evolutionary-
   biology-diet-destiny.html.

Nine: The Two Dianas

Beresford-Kroeger, Diana. *Arboretum America: A Philosophy of the Forest.*
   Ann Arbor: University of Michigan Press, 2003.
———. *The Global Forest.* New York: Viking, 2010.
Ostlund, Lars, Lisa Ahlberg, Olle Zachrisson, Ingela Bergman, and
   Steve Arno. "Bark-Peeling, Food Stress and Tree Spirits: The Use
   of Pine Inner Bark for Food in Scandinavia and North America."
   Journal of Ethnobiology 29 (2009).
Six, Diana L. "Climate Change and Mutualism." *Nature,*
   7 October 2009.

Ten: The Parable of the Worm

Dickison, R.B.B., Margaret J. Haggis, R.C. Rainey, and L.M.D. Burns.
   "Spruce Budworm Moth Flight and Storms: Further Studies Using
   Aircraft and Radar." *Journal of Climate and Applied Meteorology*
   25 (1986).
Gunderson, Lance, and C.S. Holling, eds. *Panarchy: Understanding
   Transformations in Human and Natural Systems.* Washington, D.C.:
   Island Press, 2002.

Holling, C.S. "Engineering Resilience versus Ecological Resilience."
In National Academy of Engineering, *Engineering within Ecological
Constraints*. Washington, D.C.: National Academy Press, 1996.
———. "From Complex Regions to Complex Worlds." *Ecology and
Society* 9 (2004).
———. "The Functional Response of Invertebrate Predators to Prey
Density." *Memoirs of the Entomological Society of Canada* 48 (1966).
Holling, C.S., and Gary Meffe. "Command and Control and the
Pathology of Natural Resource Management." *Conservation
Biology* 10 (1996).
Holling, C. S., G.B. Dantzig, C. Baskerville, D.D. Jones, and W.C. Clark.
"A Case Study of Forest Ecosystem/Pest Management." *Proceedings:
International Canadian Conference on Applied Systems Analysis*, 1975.
Available at www.iiasa.ac.at/Admin/PUB/Documents/
WP-75-060.pdf.
Homer-Dixon, Thomas. "Our Panarchic Future." *Worldwatch Magazine*,
13 February 2009. Available at www.worldwatch.org/node/6008.
Irland, Lloyd. "Pulpwood, Pesticides, and People: Controlling Spruce
Budworm in Northeastern North America." *Environmental
Management* 4 (1980).
Ludwig, Donald, Brian Walker, and Crawford S. Holling. "Sustainability,
Stability and Resilience." *Ecology and Society* 1 (1997).
Morris, R.F. "The Dynamics of Epidemic Spruce Budworm Populations."
Memoirs of the Entomological Society of Canada 31 (1963).
Prebble, M.L., ed. *Aerial Control of Forest Insects in Canada*. Ottawa:
Department of the Environment, 1975.
Taleb, Nassim. *The Black Swan: The Impact of the Highly Improbable*. 2nd
edition. New York: Random House, 2010.

Acknowledgments

· · · · ·

THE SCOPE and breadth of any insect empire, as well as that of a conifer forest, can be daunting without expert guides. I was fortunate to have some of the best in bark beetle country over the last four years. Allan Carroll, an insect ecologist, first introduced me to the workings of the pine beetle during a "beetle tour" of the interior of British Columbia. His science, recollections, and observations helped shape this book. On a trip to the Lubrecht Experimental Forest near Missoula, Montana, Diana Six, one feisty entomologist, patiently challenged everything I thought I knew about fungi. Diana Beresford-Kroeger, one of Canada's national treasures, taught me much about trees and the darkening state of the global forest. An entomologist, a musician, and a pool hustler—Richard Hofstetter, David Dunn, and Reagan McGuire—reminded me how real things get done in this world: through democratic and innovative collaboration. They introduced me to phloem sandwiches and some remarkable science on the acoustic ability of bark beetles. Dunn's CD *The Sound of Light in Trees* remains a subversive revelation.

From Alaska, Ed Berg, Skeeter Werner, and Ed Holsten shared many magical stories. Berg's incisive field notes made the retelling of the Alaska storm a compelling exercise.

In British Columbia, the heart of the pine beetle storm, many citizens showed me the wake of the empire. Dave Jorgenson, a third-generation logger, explained the finer points of logging beetle wood; Annerose Georgeson drove me through the beetle kill on the Kluskus road; Dave Neads eloquently debunked beetle myths; and Randy Saugstad talked about watershed destruction. Ben Parfitt, a Victoria-based journalist and forestry expert, patiently answered many queries. Ernie Graham and Claude Paquet offered honest views from the logging trenches. A variety of beetle scholars at the University of Northern British Columbia provided helpful commentary, including Peter Jackson, Staffan Lindgren, Art Fredeen, and Kathy Lewis. The communities of Prince George, Williams Lake, Mackenzie, Vanderhoof, and Quesnel offered examples of sharp ways of thinking about the fragility of the province's logging industry.

In Montana, Louisa Willcox, a powerful wildlife advocate, marched me up the Gallatins into the declining world of the whitebark pine, thanks to a timely introduction by Josh Mogerman of the National Resources Defense Council. While recovering from a skiing accident, Jesse Logan recounted beetle politics under the George Bush administration in Emigrant, Montana. Bruce Gordon of EcoFlight, along with Wally Macfarlane and Willie Kern, gave me a sobering bird's-eye view of dying forests outside of Yellowstone National Park.

Mike Wagner and organizers of the 2010 Western Forest Insect Work conference in Flagstaff, Arizona, welcomed me with open arms, as if I were some long-lost beetle chronicler. I am grateful for their generosity and hospitality. Amid much insect banter, John Borden and Ken Raffa both found time to talk about their favorite subject: bark beetles. In Edmonton, George Ball opened the doors to the E.H Strickland Entomological Museum and much wonder.

Charlotte Stenberg, a Vancouver-based translator, expertly rendered portions of Johann Gmelin's *Abhandlung über die Wurmtroknis* ("Treatise on the Worm Dryness") into English.

While I got lost in the arboreal world of bark beetles, my family generally wondered what the hell I was writing about. So did I. The book went through several changes before the significance of the beetle empire hit me with the force of a Marine landing. It took a while for the scolytids to wipe out all sorts of traditional prejudices, but in the end Mother Nature's forest engineers prevailed.

My gratitude and love go once again to my wife, Doreen Docherty, who tended to our familial forest while I got lost in collapsing ecosystems. Aidan Nikiforuk confirmed the title of this book, while Torin and Keegan grew like ponderosa pine.

Barbara Pulling, my keen editor, cleaned away all the wordy frass and helped structure an imperial narrative. My thanks again to the dependable and competent crew at Greystone (Emiko Morita, Carra Simpson, and Rob Sanders), who helped transform this book idea into something much richer than a weird science story. A generous grant from the Canada Council kept the wolf from the door. Debra Ward, the daughter of a park warden, inked the fine tree silhouettes.

Smarter and wiser writers blessed me with a rural vision: Wendell Berry and Wallace Stegner. Nassim Nicholas Taleb and Buzz Holling added critical perspective on the revenge of small things in complex economies.

May we learn from the way of the beetle.

Readers can contact me at: Andrew@andrewnikiforuk.com

Index

. . . . .

The David Suzuki Foundation

The David Suzuki Foundation works through science and education to protect the diversity of nature and our quality of life, now and for the future.

With a goal of achieving sustainability within a generation, the Foundation collaborates with scientists, business and industry, academia, government, and non-governmental organizations. We seek the best research to provide innovative solutions that will help build a clean, competitive economy that does not threaten the natural services that support all life.

The Foundation is a federally registered independent charity that is supported with the help of over 50,000 individual donors across Canada and around the world.

We invite you to become a member. For more information on how you can support our work, please contact us:

The David Suzuki Foundation
219–2211 West 4th Avenue
Vancouver, BC
Canada V6K 4S2
www.davidsuzuki.org
contact@davidsuzuki.org
Tel: 604-732-4228
Fax: 604-732-0752

Checks can be made payable to The David Suzuki Foundation. All donations are tax-deductible.

Canadian charitable registration: (BN) 12775 6716 RR0001
U.S. charitable registration: #94-3204049

DOREEN DOCHERTY

ANDREW NIKIFORUK is an award-winning Canadian journalist who has written about education, economics, and the environment for the last two decades. His books include *Pandemonium; Saboteurs: Wiebo Ludwig's War against Oil*, which won the Governor General's Literary Award for Non-Fiction; and *The Fourth Horseman: A Short History of Plagues, Scourges and Emerging Viruses*. His book *Tar Sands*, published to wide acclaim, won the Rachel Carson Environment Book Award and became a national bestseller. He lives in Calgary, Alberta.

MIX
Paper from
responsible sources
FSC® C016245
www.fsc.org